Health & Safety

at Work Essentials

Mary Duncan
Finbar Cahill
Penny Heighway
of Henmans Solicitors

Health & Safety at Work Essentials
by Mary Duncan, Finbar Cahill and Penny Heighway

Published by
Lawpack Publishing Limited
76–89 Alscot Road
London SE1 3AW

www.lawpack.co.uk

First edition 2002
Second edition 2003
Third edition 2004
Fourth edition 2005
© 2005 Henmans Solicitors

ISBN: 1 904053 77 7

The right of Mary Duncan, Finbar Cahill and Penny Heighway to be identified as the authors of this work has been asserted by them in accordance with the Copyright, Designs and Patents Act 1988.

Crown Copyright material is reproduced with the permission of the Controller of HMSO and the Queen's Printer for Scotland.

Every effort has been made to trace the copyright holders. The publisher apologises for any unintentional omissions and would be pleased in such cases to place an acknowledgment in future editions of the book.

Exclusion of Liability and Disclaimer

While every effort has been made to ensure that this Lawpack publication provides accurate and expert guidance, it is impossible to predict all the circumstances in which it may be used. Accordingly, neither the publisher, author, retailer, nor any other suppliers shall be liable to any person or entity with respect to any loss or damage caused or alleged to be caused by the information contained in or omitted from this Lawpack publication.

For convenience (and for no other reason) 'him', 'he' and 'his' have been used throughout and should be read to include 'her', 'she' and 'her'.

Contents

About the authors

Henmans Solicitors (www.henmans.co.uk) has a long-established personal injury practice, handling claims nationally for both claimants and defendants, from an Oxford base. Henmans is recognised as a leading personal injury firm in the legal directories. A large specialist team deals with cases, from serious brain injuries and fatal accidents to minor road traffic accidents. The bulk of the work undertaken relates to medium to large claims, arising out of accidents at work, road traffic accidents and clinical negligence.

Three members of the team have combined forces to produce this 'one stop' book for anyone responsible for health and safety issues in the workplace.

Mary Duncan is head of Henmans' personal injury and clinical negligence department. She specialises in major personal injury litigation for claimants and defendants, clinical negligence for claimants and professional indemnity work for defendants. She is a member of the Law Society's Personal Injury Panel and the Action against Medical Accidents (AvMA) Panel. She is also co-author of *Fatal Accident Litigation*, published by Fourmat Publishing (1993), and has written a chapter in *Trauma Care*, published by Butterworths (2000).

Finbar Cahill is a partner in Henmans' personal injury department. He acts predominantly, although not exclusively, for insurers. He specialises in personal injury claims, particularly serious and fatal accident claims, brain injury cases and employer's liability including HSE prosecutions. He is a member of the Law Society's Personal Injury Panel.

Penny Heighway is a partner in Henmans' personal injury department. She handles all types of personal injury claims, specialising in accidents at work, industrial injury cases and professional indemnity work for defendants. She is a member of the Law Society's Personal Injury Panel. She is also editor of the book *Trip & Slip: A Guide to Making Personal Injury Claims*, also published by Lawpack.

Foreword

In the last 30 years, the obligations upon and duties of an employer, in the area of health and safety in the workplace, have mushroomed. The combination of European legal influences and a more litigious society has led to a rise in legal claims which has left many businesses struggling to secure compulsory employer's liability insurance. Without such insurance a business is no longer able to trade.

Business owners are experts in business, not in insurance or health and safety law. This book is an attempt to explain, in a straightforward way, the legal requirements that an employer needs to comply with in this increasingly complex area.

We would like to thank counsel, Charles Brown and Judith Ayling of 39 Essex Street, for their consideration and time in reviewing the content of this book and for offering many useful comments. We would also like to thank Gemma Morris, Liz Stevens and Amanda Shearer for their assistance. The writers emphasise that any mistakes are their own and ask the reader to note that the law is correct as at 1 January 2005.

Mary Duncan
Finbar Cahill
Penny Heighway

CHAPTER 1

General information

Background

At least one person is killed and more than 6,000 are injured at work every working day in the UK. Every year, three-quarters of a million people take time off work due to work-related illness and, as a result, about 30 million work days are lost.

In the UK, absenteeism is costing employers £13 billion a year. Small companies, where the absence of one or two employees can make a huge difference, are the hardest hit.

Employers must have insurance cover for their employees' work-related injuries and ill-health. They also need insurance for accidents involving vehicles and possibly public liability and buildings insurance. However, various changes in the last 30 years have made employer's liability insurance an unattractive class of business for insurers. With mounting losses, the only companies they want on their books are those with a genuine commitment to health and safety. Many companies are being forced out of business, unable to trade without insurance cover or because of soaring premiums. Those who have difficulty getting renewal terms have often shown a disregard for health and safety issues in the past.

However, insurance policies only cover a small proportion of the costs of accidents. Costs not covered by insurance can include the following:

- Sick pay
- Damage or loss of products and raw materials
- Repairs to the plant and equipment
- Overtime working and temporary labour
- Production delays
- Investigation time
- Criminal fines

On a more positive note, managing the well-being of employees does offer the employer a means of reducing costs and motivating the workforce.

The law states that it is an employer's duty to ensure that the risks to health and safety are properly controlled and that there is the correct provision for the employees' welfare. In certain circumstances, this responsibility extends to the self-employed and to the protection of the health and safety of other people who may be affected by your business.

It is not always easy for business owners to find out exactly what is expected of them. It is hoped that this book, which is aimed towards the small- to medium-sized employer in all areas of employment from office work to heavy industry, will help you navigate the important aspects of health and safety law. This will, hopefully, ensure that not only those on your premises are protected but that you are also protected from breaking the law.

The Health and Safety Executive

In the UK, the Health and Safety Executive (HSE) is the main body responsible for enforcing legislation and providing guidance on health and safety in the workplace.

The HSE comprises over 4,000 staff. As well as inspectors, there are policy, medical and scientific advisers.

In addition to the HSE, more than 400 local authorities are used to enforce the relevant law. Your local fire brigade also has certain responsibilities.

The shared responsibility for enforcing health and safety is split by the type of business. The enforcing bodies dealing with each type of business are listed below:

- **The HSE** covers offices, factories, building sites, mines and quarries, fairgrounds, railways, chemical plants, offshore and nuclear installations, schools and hospitals.
- **The local authority** enforcement officers cover retailing, warehouses, most offices, hotel and catering, sports, leisure and consumer services.
- **The local fire brigade** deals with fire safety regulations.

The law grants the enforcing bodies a wide range of powers. Amongst others, they have the right to do the following:

- Enter the premises at any reasonable time.
- Carry out examinations and investigations, take measurements, photographs and samples, take possession of articles and arrange for them to be dismantled or tested.
- Require information, take statements from people and inspect copy documents.
- Issue improvement and prohibition notices and prosecute people in companies.

We will deal more fully with the role of the HSE in detail in chapter 4, sections 2 and 5.

Legal requirements

Breaches of health and safety law can incur both criminal and civil liability.

Criminal liability

A crime is an offence against the state. In health and safety law, criminal liability refers to the duties and responsibilities under statute, primarily the Health and Safety at Work Act 1974 (also known as the 'HSW Act'). There are also several sets of regulations. These implement European legislation and provide the minimum health and safety requirements expected of an employer.

Both the HSW Act and the regulations are written in complicated legal jargon, so in order to help explain them in a language that is easier to understand approved codes of practice (ACOP) and guidance notes are issued regularly. The ACOP have special legal status similar to that of the Highway Code.

The penalties that can be imposed by the criminal courts include fines and imprisonment.

Civil liability

A civil action involves negligence and/or breach of legislation. In addition to the HSW Act and the regulations, various cases have also established legal principles which employers are obliged to follow. In a civil action a 'claimant' (the wronged person) sues a 'defendant' (the wrongdoer) for a remedy – usually financial compensation.

Breach of a duty imposed on an employer by the regulations does not, however, necessarily give a person a right of action in civil proceedings against an employer. Legal advice from a Law Society recommended solicitor should always be sought.

The Health and Safety at Work Act 1974 and regulations

The HSW Act requires the business owner to carry out actions that are reasonable and practicable in order to protect the workforce. The Act covers not only those who work full-time but also part-timers, casual workers and outside contractors. It also includes those who may use your business premises or equipment, which may include visitors and delivery drivers.

The regulations tend to relate to specific issues or risks in the workplace. We will deal with those risks commonly found in the workplace in detail in chapter 3.

It is, of course, impossible to make your premises 100 per cent safe and risk free, and both the HSW Act and regulations recognise this. As a general rule, to prosecute criminal proceedings successfully, the HSE must prove beyond reasonable doubt that the business owner failed to act as a reasonable employer in all the circumstances and that the injuries were caused as a direct result of this failure.

Six sets of regulations came into force in 1992 (but some of them have since been updated and are also subject to more recent amendments). They are often referred to as the 'Six Pack'. These are as follows:

- The Management of Health and Safety at Work Regulations 1999

- The Workplace (Health, Safety and Welfare) Regulations 1992

- The Provision and Use of Work Equipment Regulations 1998 (PUWER)

- The Manual Handling Operations Regulations 1992

- The Health and Safety (Display Screen Equipment) Regulations 1992 (DSE)

- The Personal Protective Equipment at Work Regulations 1992 (PPE)

There are other additional regulations. The Six Pack and other main legislation are summarised below. Many are dealt with in more detail in later chapters. It is important to know the contents of the regulations so that you are aware of your obligations and duties as an employer.

The Management of Health and Safety at Work Regulations 1999

These regulations are accompanied by an approved code of practice. The duties are of a strict nature, compared with those under the HSW Act and other regulations that are qualified by the term 'so far as is practicable' or 'reasonably practicable'. Therefore you must comply with the precise terms of these regulations.

To ensure a safe workplace these regulations state that it is the employer's responsibility to do the following:

- Carry out a suitable and sufficient assessment of the risks to the health and safety of employees in order to identify hazards, evaluate the extent of risks and take appropriate action. In addition, an employer must carry out a further assessment after an accident has occurred.

- Record the arrangements for health and safety including planning, organisation, control and monitoring as well as undertake a review of the 'protective and preventative measures' in place. Every employer must appoint one or more competent persons to assist him in undertaking these tasks.

- Employ competent people to carry out the work duties. You must not employ children of under school-leaving age, unless their employment is part of an authorised scheme.

- Ensure that any necessary contacts with external services are arranged, for example, first aid and emergency medical care.

- Provide comprehensive health and safety information to all employees, including temporary employees. This includes information on risks to health and safety as identified by risk assessments, preventative and protective measures and the identity of the competent person outlined above. Be aware of communication issues such as language differences.

- Make arrangements for the safety of employees of other companies on site and provide appropriate instructions and information.

- You should note that if fellow workers are injured as a result of another worker carrying out work for which he is untrained, the business owner can be liable.

- Provide safety training at induction, or upon transfer to a new job or area, or when new equipment is introduced. (Keep training records as evidence of this.)

- Employees are required to report any shortcomings in safety arrangements to their employers.

- Following an accident, the employer is required to complete an accident investigation (and record the findings).

- Keep records that you have undertaken all of the above.

The Workplace (Health, Safety and Welfare) Regulations 1992

These state that your premises should not be, or create, a risk to your workforce. This is not restricted to the building itself but covers four main areas:

- **Working environment:** includes temperature, ventilation, lighting and emergency lighting, room dimensions and space, suitability of workstations, seating and outdoor workstations.

- **Safety:** includes safe passage for pedestrians and vehicles. Windows and skylights must be capable of being opened, closed and cleaned safely. Transparent doors and partitions must be marked and made of safe material. There should be safety devices on doors, gates and escalators and floors must be constructed safely and be free from hazards such as slipping, tripping and obstructions. Falls from heights, dangerous substances and falling objects should be prevented by guards or safety rails. Warning signs should be placed where necessary.

- **Facilities:** there must be suitable and sufficient lavatories, washing facilities and changing areas together with storage areas and lockers for clothing. There should be adequate rest facilities and areas for eating and drinking. Drinking water must also be provided.

- **Housekeeping:** your workplace must be kept clean and maintained at all times. Spillages must be cleaned up immediately and provision must be made for the removal of waste material. You must keep housekeeping, repair and maintenance records.

The Provision and Use of Work Equipment Regulations 1998 (PUWER)

These state that in the course of carrying out their jobs, workers must be provided with machinery and equipment that is safe to use. It is important to note that employers are responsible for any injury sustained by a worker as a result of defective machinery, even if the defect arose during the manufacture of the machinery.

Work equipment is broadly defined under these regulations. It includes everything from a small screwdriver to a major plant (e.g. an oil refinery) but excludes construction sites, which are dealt with in other regulations. The general duties of an employer are to:

- consider the risks in the workplace before buying new equipment;

- provide adequate training, information and instruction on the equipment (and document as much as possible);

- ensure that the equipment is properly maintained (and keep maintenance and repair logs);

- protect the user from dangerous parts of the machinery (e.g. by fitting guards and emergency stop controls);

- reduce the danger caused by specific hazards;

- ensure the stability of the equipment, provide adequate lighting and make warnings and markings clearly visible. See chapter 3 for further information.

The Manual Handling Operations Regulations 1992

The regulations require the employer to:

- assess manual handling tasks in the workplace by way of a risk assessment (and keep copies of the assessments including post-accident assessments);

- avoid manual handling in so far as is reasonably practicable;

- mechanise the task if possible;

- reduce the risk of any injury to the lowest level reasonably practicable;

- train staff in safe methods of lifting if necessary and document all the training and information given to staff. (see chapter 3, section 2 for further information);

- provide information to staff as to the weight, or heaviest side, of the load.

The Health and Safety (Display Screen Equipment) Regulations 1992 (DSE)

These regulations cover screens and workstations. They require the employer to:

- analyse the workstations to assess and reduce risks, and to keep copies of the assessments;

- plan work to ensure adequate breaks;

- provide information and training to users of equipment and keep documentation relating to this;

- consult on software design and selection (see chapter 3, section 4 for further information).

The Personal Protective Equipment at Work Regulations 1992 (PPE)

Personal protective equipment (PPE) is all equipment designed to protect against risks to health and safety and includes most types of equipment not covered in other legislation. PPE must be used only as a last resort, where another safer system of work cannot be implemented. If PPE is needed, it must be provided free by the employer. See chapter 2, section 8 for further information.

The Reporting of Injuries, Diseases and Dangerous Occurrence Regulations 1995 (RIDDOR)

The reporting of certain accidents and ill-health at work is a legal requirement. This is dealt with in detail in chapter 4, section 1.

The Control of Substances Hazardous to Health Regulations 2002 (COSHH)

These are often referred to as COSHH and they require an employer to:

- assess all the substances held in the workplace which are, or may be, hazardous to health (see chapter 3, section 5 for further information);
- ensure that all hazardous substances are labelled correctly with the appropriate signs;
- undertake separate assessments for PPE (see chapter 2, section 8 for further information);
- provide instruction and training in relation to PPE and training in general;
- conduct health surveillance if necessary.

The Noise at Work Regulations 1989

The employer is under a general duty to arrange for an adequate noise assessment and to reduce the risk of damage to the lowest level which is reasonably practicable. Special risk assessments must be undertaken. See chapter 3, section 9 for further information.

The Health and Safety (First Aid) Regulations 1981

These regulations place a general duty on the employer to make adequate first aid provision for their employees if they become injured or ill at work. See chapter 2, section 2 for further information.

The Waste Management and Environmental Protection Act 1992

This Act demands certification from an employer of all waste removed from a site. The purpose of the Act is to ensure the safe disposal of all waste in a proper and safe manner to an appropriate disposal site. This includes the effective disposal of waste which may seep through soil and enter the water table, as well as the removal of hazardous materials and liquid.

The Safety Representatives and Safety Committees Regulations 1977 and the Health and Safety (Consultation with Employees) Regulations 1996

Although it is unlikely to apply to a small business, if you do have union representation you must consult them on health and safety matters and in prescribed cases, appoint a safety representative and/or safety committee.

In any event, you must consult your employees on issues regarding changes affecting health and safety and the provision of information and training.

The Health and Safety Information for Employees Regulations 1999

These regulations require information relating to health, safety and welfare to be given to employees by means of posters or leaflets in the approved form. Copies of the posters and leaflets can be obtained from the HSE (see chapter 2, section 2 for further information.)

Other regulations

Other regulations which deal with health and safety issues include the Electricity at Work Regulations 1989, the Safety Signs Regulations 1996 and the Lifting Operations and Lifting Equipment Regulations (LOLER) 1998.

If you are a builder or at all involved in construction, you will need to look at the Construction (Health, Safety and Welfare) Regulations 1996, the Construction (Design and Management) Regulations 1994 and the Construction (Head Protection) Regulations 1989 (refer to chapter 3, section 14 for further information).

This list is not exhaustive. It is advisable to obtain full copies of the most relevant and up-to-date regulations to ensure that you comply with them. Copies can be obtained from the HSE by calling their Infoline on 0870 154 5500. It is possible to obtain some of their leaflets and publications for free.

CHAPTER 2

Starting a business
Initial considerations for the employer

Registering your business

If you are starting a commercial or industrial business and are employing people, you are legally obliged to notify your local Health and Safety Executive (HSE) inspector or local authority. Check which enforcing body is responsible in chapter 1, section 2. If your business involves manufacturing or processing or providing a service such as dry cleaning or telephone repairs, ring HSE's Infoline and they will send you a form to complete (see the Appendices for their contact details),

If your business is a shop, office, restaurant, hotel, launderette or residential home, your local authority's environmental health department will send you a form. If you are in doubt as to who to register with, telephone the HSE's Infoline.

Providing information to employees

You should be aware that certain health and safety information must be provided to your employees in order to comply with the law. This includes:

1. the Health and Safety Law poster/leaflet;

2. the location of first aiders and the first aid box;

3. the location of the accident book;

4. the health and safety policy document.

The health and safety law poster

It is a legal requirement that a Health and Safety Law poster is displayed in the workplace. A large A2 poster is published by the HSE and must be displayed prominently so that employees have the opportunity to familiarise themselves with the law on health and safety at work. Alternatively, a leaflet can be given to all workers. These posters or leaflets can be obtained directly from the HSE. Details, which need to be put on the poster, include the:

- name of an employee representative (if there is one);

- name of a management representative (which could be the employer in a small business, provided that he has the correct training);

- contact details of your enforcing authority.

First aiders

Legal considerations apart, there is clearly a case for having as many people as possible trained in basic first aid procedures. Such knowledge could be instrumental in saving the life of a person whatever the circumstances. However, the law says that you must have:

- a person in your workplace who can take charge in an emergency;

- a first aid box;

- a notice stating where the first aid box is and who the approved person is;

- a trained first aider and first aid room if your workplace gives rise to special hazards.

Once the health and safety risks associated with your particular business have been assessed (see section 3 for further information), you can consider whether you simply need to nominate a person to be responsible for first aid or whether you need to have a trained first aider. Also you will need to know what the minimum number of items in your first aid box should be, how many first aid boxes are required and where they should be located.

On the next page is a guide as to the requirements for first aiders. It is not a legal requirement. As your company grows, it is important to reassess the need for qualified first aiders.

The Health and Safety (First Aid) Regulations 1981 lay down various requirements for the provision of suitably trained first aiders in the workplace. Training courses in first aid at work are provided by training organisations, which must be approved for this purpose by the HSE. Currently, the HSE accepts courses run by the Resuscitation Council (UK), St. John Ambulance, St. Andrew's Ambulance Association and the British Red Cross.

Once a course is completed, your employee holds a valid first aid at work certificate. This must be renewed every three years. A shorter updating course can then be undertaken. However, if the course is not attended before the expiry of the three years, then the employee must completely retrain.

You will find further details of training courses and organisations in your area from your local HSE office or your local employment medical adviser.

Do remember that first aiders and people with first aid responsibility will be away from work from time to time so you need to arrange appropriate cover.

You should also be aware that the compulsory element of the employer's liability insurance does not cover litigation resulting from first aid to non-employees. However, many public liability insurance policies do cover this aspect and you may want to check your public liability insurance policy on this point.

If you have a first-aid room, it must be easily accessible to stretchers and be sign-posted.

Type of business	Number of workers	First aid provision
Retail shops, offices (accountants, solicitors, etc), public buildings, car showrooms	Fewer than 50	At least one responsible employee
	Between 50 and 100	At least one first aider
	More than 100	One additional first aider for every 100 employed
Assembly plants, warehousing units, engineering business, food production	Up to 20 employees	At least one responsible employee
	Between 20 and 100	Two first aiders
	More than 100	One additional first aider for every 100 employed
Chemical plants/ manufacturing, building/construction sites, heavy engineering, biological plants, slaughterhouses	Up to five employees	At least one responsible person
	Between five and 50	At least one first aider
	More than 50	At least one additional first aider for every 50 employed

The first aid box

All first aid containers must be identified by a white cross on a green background. You need to assess the level of health and safety risks your workers face as it is impossible to be definitive about what your first aid box should contain. For a business where there is no heightened level of risk to its workforce, here is a list of items that may be regarded as a minimum requirement:

- A first aid guidance leaflet
- Two pairs of disposable gloves
- Four sterile eye pads
- Six to eight safety pins (various sizes)
- Four sterile bandages (triangle shaped)
- Five large individually wrapped sterile wound dressings
- Two 5cm 5 5cm bandages
- Two 2.5cm 5 5cm bandages
- Ten absorbent cotton wool balls
- One zinc oxide plaster (for securing dressings)
- 15 multi-stretch plasters (various sizes)
- Pack of antiseptic wipes
- Tube of antiseptic cream
- One pair of small scissors

You should not keep tablets or medicines in the first aid box. If you use dangerous plant machinery, toxic chemicals or hazardous substances (e.g. cleaning solvents), you may require different items to be contained in your first aid kit.

In addition, you should consider whether different areas of the business premises pose a different risk to others, for example:

- How spread out are the premises?
- Are there several buildings?
- Is there more than one floor and, if so, do you need separate medical facilities on each?

Finally, if you employ people with special needs or disabilities you must consider whether these needs are catered for.

The accident book

All employees should be informed where the company's accident book is located. This could be at reception or, in larger organisations, in the first aider's room.

There are various accident record books on the market or you could maintain the record in a computer file. However you choose to document accidents, certain specific information is required:

- **Personal details:** name, age, status, job title
- **Injury details:** nature of accident and injury/disease
- **When the accident occurred:** date, time, place

The health and safety policy document

If you employ five or more employees, you must have a written health and safety policy and bring it to the attention of your employees.

According to section 2(3) of the Health and Safety at Work Act, the policy should:

- set out the employer's general policy with respect to health and safety.;
- describe the organisation and arrangements for carrying out that policy;
- be reviewed as often as appropriate (e.g. annually).

Having your own policy statement is very important. Health and safety inspectors often ask to see the policy statement during inspection visits. A policy statement demonstrates that a business is committed to planning and managing health and safety. Before finalising your policy statement, it is a good idea to consult your workers. This helps to ensure that all the necessary systems and procedures are in place. It should be clear from your statement who is responsible for the different areas of the health and safety requirements. Do remember that after this policy has been written, discussed and agreed with the workers, it should be implemented!

Preparing a health and safety policy document

Often business owners find it difficult to formulate and document their policy. The following guidance therefore may be helpful in providing a template for the small business owner.

It should be in written form and signed and dated by the owner, occupier or person having control of the business.

Depending on the type of business, the policy statement can have up to 11 different sections. We will cover all of them and leave it to you to select those appropriate to your business.

Section 1: A general statement of intent

Here you describe in broad terms your company's philosophy in relation to health and safety. You need to include a statement saying that you will provide adequate consultation with your employees and ensure the prevention of accidents and work-related ill-health, the safe handling of toxic substances and the maintenance of safe plant machinery. In addition, you need to state that you will review and revise your policy as necessary, particularly as the business changes in nature and size. This section should be signed and dated.

Section 2: Organisation

This section deals with people and their duties. Here you should list four areas of responsibility. Firstly, as the business owner, you have overall responsibility for health and safety. Your name goes first. Secondly, you should appoint an employee who is responsible for the day-to-day implementation of the policy. Thirdly, there may be instances where different people have different specific areas of responsibility for ensuring that health and safety standards are maintained. These people could make up a health and safety committee who meet regularly (e.g. quarterly) for the purpose of reviewing health and safety procedures and making policy decisions. These people should be named here. Finally, it should be stated that 'all employees have a duty to take reasonable care of their own health and safety and that of others who may be affected by their acts or omissions and to co-operate with their employer, so far as is reasonably necessary, to ensure compliance with the related statutory requirements'.

Section 3: Health and safety risks arising from workplace activities

Here you will document those responsible for assessing the risks in the workplace (see section 4 in this chapter). You will need to note down the person undertaking the risk assessment, to whom the results will be reported and those responsible for any action that needs to be taken. The actual risk assessments can be attached and you will also want to include a period of time when assessments will be reviewed.

Section 4: Consultation with employees

Even if you do not have a trade union, you must still consult your employees. If this is undertaken through a nominated representative, their name needs to be documented here. If you have a union, you must consult them.

Section 5: Safe plant and equipment

This section is used to identify all the plant machinery that you possess and the people responsible for its maintenance. You should list who is responsible for identifying when maintenance is needed, who draws up the maintenance procedures, who to report problems to and who purchases new equipment. A maintenance logbook is recommended. Don't forget simple tasks such as the routine inspection of plugs and cables for loose connections and faults. It should be stated that under no circumstances must an unqualified member of staff attempt any electrical repair or maintenance.

Section 6: Safe handling and use of substances

If you use any type of hazardous substances (e.g. photocopying toner, Tippex, thinners, bleach, etc), these need to be identified here and the special risks to health assessed. Whoever identifies the hazards and is responsible for the detailed assessment which is required under the Control of Substances Hazardous to Health Regulations 2002 (COSHH) must be noted (chapter 3, section 5 deals with this area in detail).

Also list the assessment review period and the names of others who are involved in the implementation of COSHH.

Section 7: Display safety information

All business premises must display the health and safety law poster (see page 14). Here you can document where it is displayed, along with details of where the employee health and safety information is kept.

Section 8: Job training/induction/advice and consultancy

All new employees need to be given a comprehensive health and safety induction programme and the person responsible for this training should be included here. Training new personnel in tasks specific to their jobs is also important and those workers responsible for this should be named. Training records should be kept and their whereabouts noted.

If any outside agencies are used to assist with health and safety advice and training, they should also be listed here.

Section 9: First aid/accident procedure/work-related ill-health

Document here where the first aid box(es) and equipment are stored, who the people responsible for administering first aid are, who keeps the records and, in the event of a serious accident, the person responsible for reporting to the enforcing authority what has happened.

All accidents at work should be recorded and kept in an accident record book. The location of the book should be noted here.

Section 10: Safe working practices

You should nominate an employee whose job it is to monitor and check regularly your safe working practices. They should also be responsible for investigating accidents and work-related sickness.

Investigating the reasons why accidents have occurred is a useful exercise that enables you to tighten procedures and hopefully prevent a recurrence. Descriptions of specific safe practices should be listed here, for example, the safe use of photocopiers, manual handling of heavy loads, the use of VDUs, etc.

Section 11: Emergency procedures: fire and evacuation

Regularly checking fire and emergency exits, escape routes and fire alarms is vital. Appoint a responsible person to carry out this procedure and name them. Put in specific dates/times when alarms, escape routes and fire extinguishers will be tested, along with the emergency evacuation procedures. There should also be random testing of the emergency evacuation procedures. Talk with your local fire service to get further guidance on your obligations. This issue is dealt with in more detail in section 3 of this chapter.

More specific guidance

This checklist is only a guide to what you should consider including in your company's health and safety policy document. It is by no means exhaustive and different company circumstances will determine what needs to be included and what can be safely excluded.

In order to help the small business with the preparation of its own policy, the HSE has produced an excellent health and safety publication which employers can use to comply with the law. It's free and is called *Stating Your Business* (INDG324). For those businesses with internet facilities, you can download the entire document from the internet at www.hse.gov.uk/pubns/indg324.pdf. (Refer to the Appendices for other useful addresses and leaflets produced by the HSE.)

Emergency procedures

The Management of Health and Safety at Work Regulations 1999 require employers to establish 'procedures for serious and imminent danger and for danger areas'. The risk assessments required under the regulations (dealt with in detail in the next section) should identify any significant risks which arise out of the specific workplace. These may include, for example, the potential for a major escalating fire, explosion, building collapse, pollution incident and/or bomb threat.

Fundamentally, the question must be asked, 'What are the worst possible types of incident that could arise from the work undertaken in my workplace?' Once these major risks which may result in serious and or imminent danger have been identified, a formal emergency procedure must be produced.

The procedures should set out the role, identity and responsibilities of the competent persons nominated to implement the action. These procedures should normally be written down, clearly setting out the limits of action to be taken by all employees. In particular, work should not be resumed after an emergency if serious danger remains. You should always consult the emergency authorities if in doubt.

In shared workplaces, separate emergency procedures should take account of others in the workplace and should, as far as is appropriate, be co-ordinated.

Fire prevention

Every year, losses to business through fire are substantial. In order to prevent death and human injury, damage to property and the consequent losses, it is essential that all your workers are familiar with the causes of fire, the fire protection procedures and the dangers associated with flammable substances. Specific training should be given on this important subject and specific information can be obtained from various external agencies. (See the Appendices for suggestions.) Training can include watching videos which can quickly and easily show employees how to cope in an emergency.

Fire appliances

Different fire appliances should be used depending on the type of fire to be tackled. For example, to douse solid materials, usually organic with glowing embers, the appropriate extinguishers are water, foam, dry powder, vaporising liquid or carbon dioxide. For liquid fires, miscible (i.e. which can be mixed) with water, the appropriate extinguishers are water, foam, carbon dioxide and/or dry powder. For liquid immiscible with water, use foam, dry power, carbon dioxide or vaporising liquid. Fire appliances introduced before 1 January 1997 are colour coded as follows:

Extinguisher	Colour code
Water	Red
Foam	Cream
Carbon dioxide	Black
Dry chemical powder	Blue
Vaporising liquid	Green

Since 1 January 1997, all fire extinguishers are (rather unhelpfully) painted red. This is due to a European Directive.

Existing extinguishers do not have to be replaced until they reach the end of their useful life. The company responsible for the maintenance of the extinguishers should make sure that, whenever possible, the two standards are not mixed to avoid confusion.

To ease the transition, British Standards allows for a coloured band at the top of the extinguisher. This is only a recommendation and throughout the UK individual manufacturers have confused matters further by colour coding in different ways! The new extinguishers have icons, indicating what type of fire they can be used on, as follows:

Classification of fire	Typical fuel	Classification icon
Class A Carbonaceous fires	• Wood • Paper • Cloth • Plastic • Rubber	

Classification of fire	Typical fuel	Classification icon
Class B Liquid fires	• Paraffin • Petrol • Types of adhesives • Types of paint • Types of spirit • Thinners	
Class C Gas fires	• Domestic gas • Butane • Acetylene • Methane	
Fires involving electricity (not a class)	Equipment: • Electric heaters • Office equipment • Electric switchboards • Electric motors • Televisions • Photocopiers • Electrically operated equipment	

	Extinguisher			
Classification of fire	Water	Carbon dioxide	Powder	Foam
A – Carbonaceous fires	Yes	Yes	Yes	Yes
B – Liquid fires	No	Yes	Yes	Yes
C – Gas fires	No	No	Yes	No
Fires involving electricity	No	Yes	Yes	No

Legal requirements

The legal requirements are outlined in the Fire Precautions Act 1971 (FPA), the Fire Precautions (Workplace) Regulations 1997 and the Management of Health and Safety at Work Regulations 1992 (as amended). They include the following:

- The FPA requires a fire certificate for premises with more than 20 employees on the ground floor or more than ten employees on other floors. The certificate is usually issued by the fire brigade.

- A means of escape in the case of fire (this excludes lifts, escalators and revolving doors). There are specific requirements as to the travel distance between any point in a building and the exit and the number of exits necessary. Specific guidance should be sought from the fire authority.

- Fire instructions must be clearly displayed, advising occupants of what action they must take on hearing the fire alarm or discovering a fire. The instructions must be displayed at prominent points in the building.

- Fire drills should result in a total evacuation of the building and should take place at least annually. Fire wardens or other designated people should take a head count on evacuation. Assembly points should be clearly identified and all persons, including visitors, should be made aware of their specific assembly point.

- The fire alarm should be sounded weekly so that all occupants of the premises are familiar with its sound.

- A fire risk assessment must be carried out. You must consider all of your employees and all other people who may be affected by a fire in the workplace. You are required to make detailed provision for any disabled people, or people with special needs, who use, or may be present on, your premises (see the following page for more information on risk assessments. However, these must be recorded if you employ more than five people). The risk assessment will help you decide the nature and extent of the fire precautions that you need to provide.

- You must establish a suitable means of contacting the emergency services and ensure that they can be called easily.

Criminal breaches of the Fire Precautions (Workplace) Regulations 1997 are punishable by a fine, or by a fine and imprisonment. Anyone who fails, intentionally or recklessly, to comply with the regulations and puts their employees at serious risk commits a criminal offence.

To sum up, staff should receive regular instruction in the selection and correct use of fire appliances. However, the first priority must always be to evacuate a building in the event of a fire.

Risk assessments

As previously mentioned in chapter 1, you are under a legal obligation as an employer to ensure that the premises, systems of work and equipment do not pose a risk to your employees. This is dealt with in full in the Management of Health and Safety at Work Regulations 1999.

The law requires that employers ensure their employees' safety by initially undertaking a risk assessment. Do not be alarmed by the term 'risk assessment'. It is all about common sense and doing the things described in this section.

All employers are legally required to assess the risks in the workplace. If you have more than five employees, you are legally required to keep a written record of the risk assessment.

But what exactly is a risk assessment, what does it entail and what is the best way for a small business owner to go about meeting his obligations when there may be limited time and money?

What are they?

Risk

A risk is a chance, be it high or low, that someone may be harmed by a hazard. A risk assessment is a careful examination of what could cause harm to people in a particular workplace.

Hazard

A hazard means anything that can cause harm, for example, working at great heights or with chemicals.

Assessment

It needs to be decided whether a hazard is significant and whether it has been covered by satisfactory precautions so that the risk is greatly reduced. If it has not, it needs to be decided what precautions should be taken.

What do they entail?

In drawing up an initial outline assessment, the guiding principle is to keep things simple. In an office or commercial operation, there should not be many dangerous items or situations. It is just a combination of observation and common sense.

It is the small, often overlooked, hazards which you probably need to concentrate on. If you are a reputable employer, you will already have ensured that potentially harmful machinery is correctly guarded and your workforce has the necessary protective clothing (e.g. boots, gloves, eye protection, etc). Instead, you may have to look at the low beam at the entrance to your building or the steep inside staircase that you need to make people aware of. Taking the correct precautions may simply mean a well-placed sign. See section 9 for example signs.

If yours is a small business, then an assessment can be undertaken within the business. A larger company may wish to ask a safety representative or safety officer to assist. You could employ a private company to undertake the risk assessment on your behalf. Insurance companies will also carry out surveys to make sure that the companies they insure are up to standard. Small businesses should not view these as potential problems but effectively as free advice which can help improve things without running up professional fees.

But remember, the employer is ultimately responsible for ensuring that the risk assessment is adequately undertaken.

How to undertake an effective risk assessment

In order to assess the risks in the workplace, the following process should be undertaken.

Stage 1: Look for hazards

Spend time, ideally with a colleague, walking around your workplace with a critical eye. Think about what could go wrong at each stage of what you do. Identify and document all areas that you think may be hazardous and the type of hazards involved. If your workplace is an office, then your list will probably be short; if your workplace is a factory, it will be considerably longer. List everything, including the low beam at your entrance or the awkward staircase.

Discuss with your workforce any concerns they have and, if legitimate, add them to the list. Do not forget your own experience. Have you had any accidents? If so, document them.

Whilst it is important to document every potential hazard, you should concentrate on the ones that are significant and avoid those that are trifling. Check your accident record book to see if there are any types of injuries or accidents that recur regularly.

Stage 2: Assess who may be harmed and how

If it is the low beam outside your building, then everyone who enters is at risk. Not only are your employees in danger of hurting themselves but your visitors are also. You would not want the embarrassment of an important customer being injured whilst visiting you. New employees, who are unfamiliar with your office/shop floor/warehouse layout, will need particular attention. Likewise any person doing work experience, temporary staff or those with special needs.

Stage 3: Consider the risk and decide whether the precautions in place are adequate or whether more could be done in taking action

Ask yourself:

1. Can I get rid of the hazard altogether?
2. If not, how can I control the risk so that harm is unlikely?

Think about the following:

1. What is the worst result? For example, is it a cut finger, suffering from asbestosis or a death?

2. How likely is it to happen?

3. How many people would be affected if things did go wrong?

You should now know what the main risks in your workplace are. However, do refer to chapter 3, which covers specific risks, just to make sure that you have covered all eventualities.

If you discover a potential hazard that may endanger someone's safety and you regard it as significant, you then need to decide how best to reduce the risk. Ideally, you should eliminate the risk altogether by getting rid of the hazard but often the situation is not as clear-cut as this.

For the most part, the law requires you to do all that is reasonably practicable to ensure your place of work is safe. At the very least, you need to do everything necessary to meet the legal safety requirements and industry safety standards. If you cannot eliminate the hazard, look at ways of controlling it, for example, a non-slip mat on slippery entrance steps. Again, refer to chapter 3 on specific risks which may help you to take the correct precautions.

There may be occasions where your office is in a building occupied by other business tenants. If this is the case, then you should consult with them on any hazards that affect everyone. For example, fire escapes, reception areas, washing/toilet facilities may be areas of joint responsibility. Also, you should discuss any concerns with your landlord where the hazard relates specifically to the building and any contents that fall under the tenancy agreement.

In many instances it will be difficult to eliminate the risk and in implementing controls your main aim will be to make the risk as small as possible.

Stage 4: Document your findings

If you employ five or more people you must record the significant findings of your assessment.

You need to cover:

- the areas you have investigated;

- any hazards you have found and whether they were significant;
- the conclusions you came to;
- the controls put in place to reduce the risks of significant hazards.

You may find it useful to include this entire document in your own health and safety policy document, which we dealt with on page 18.

Stage 5: Review the assessment and revise it as necessary

Do review the assessment, particularly following an accident at work or if new machines or substances are brought onto the site.

As your business grows, you will employ more people and invest in further, and perhaps more, sophisticated equipment. When your business grows, there will also be a growth in hazards. Remember to review your assessment regularly.

Simple Generic Risk Assessment							
Premises/Work area.........							
Assessment carried out by...........							
Date of Assessment..............							
Hazard Identified	Who Is At Risk	Effect of Hazard	Existing Controls	Action Required	By When	Review Date

Example risk assessment form

Insurance provisions

Most employers are required by law to insure against liability for personal injury or disease to their employees arising out of their employment. Recognised psychiatric or mental illness falls within the scope of personal injury. The Employer's Liability (Compulsory Insurance) Act 1969 provides that each employer should have at least a minimum level of insurance cover (£5,000,000) against any such claims. The Act does not grant an employee an automatic right to compensation. The purpose of the Act is to ensure that when an employee is successful in a civil claim, the employer can, through his insurer, pay the compensation that is due.

It is a legal requirement that a copy of the insurance certificate is displayed where your employees can easily read it.

You should be aware that people who are normally thought of as self-employed may be considered employees for the purpose of employer's liability insurance. Whether insurance is required will depend upon the terms of the contract with the person. A contract may be spoken, written or implied and there are no hard and fast rules as to who counts as an employee. However, in general you may need employer's liability insurance for someone for whom you:

- deduct National Insurance and income tax from the money you pay them;
- control where and when they work and how they do it;
- supply the majority of materials and equipment;
- possibly share your profits, either through commission, performance pay or shares.

In general, you may not need employer's liability insurance for people who work for you if they:

- do not work exclusively for you (i.e. they work as an independent contractor);
- supply most of the materials and equipment;
- are clearly in business for their own personal benefit.

To sum up, an employer needs employer's liability insurance unless they have people working for them who are not considered 'employees' or who are exempt.

Exempt employees are very restricted but can include those in a family business (i.e. if your employers are closely related to you (as husband, wife, father, mother, grandfather, etc)). However, this exemption does not apply to family businesses which are incorporated as limited companies.

If you are in any doubt as to what is required, you should seek legal advice or consult an insurance broker. You can get a list of authorised insurers via the internet at the Financial Service Authority's website (www.fsa.gov.uk) or by telephoning them on 0845 606 1234 – see the Appendices for their contact details.

Every business must retain copies of its out-of-date insurance policies for 40 years. This does seem like a very long period of time but it is necessary because the insurance covers diseases, not just accidents, which may take many years to manifest, for example, mesothelioma (see the Appendices for a list of possible work-related diseases). You can keep copies electronically, if more convenient, but they must be available to health and safety inspectors when requested.

There are considerable fines for businesses which attempt to avoid their responsibility for providing health and safety insurance cover for their workforce. The courts do take a particularly dim view of companies without insurance and fines of up to £2,500 per day can be imposed. Even if you simply fail to display your insurance certificate, you may find yourself having to pay a £1,000 penalty.

Your obligations to the insurance company

Your insurance policy is an agreement between you and your insurer concerning the circumstances in which they will pay compensation. However, your insurer cannot refuse to pay compensation to a claimant because you:

- have not provided reasonable protection for your employees against injury or disease or failed to meet a legal requirement;
- are unable to provide particular information to your insurers;

- have done something, or not done something, contrary to their instruction.

However, the policy may enable the insurer to sue you to reclaim the cost of the compensation if you do not co-operate or if you behave unreasonably. It is important to notify your insurer very promptly in the case of any accident at work, a potential claim or if the HSE notify you of a potential prosecution.

Employees abroad

Employer's liability insurance is not required to cover employees who are based abroad. However, do check the law in the country where they are based and find out if it requires you to take out insurance or carry out any other measures to protect your employees.

If any employee is normally based abroad but spends more than 14 days continuously in Great Britain or more than seven days on an offshore installation, then employer's liability insurance will be required under English law.

Other insurance policies

Any injuries or illnesses relating to motor accidents, which occur whilst your employees are working for you, may be covered separately by motor insurance. Public liability insurance covers employers for claims made by members of the public, but not for claims made by your employees. See chapter 4, section 1 for details of your obligations to report an accident to your insurer.

Recruitment and training

Your workforce is arguably the most important asset of your business. It is important to identify areas of work which place particular physical or mental demands on people. Where these demands cannot be reduced, it is important to select the right people to meet the demands. For some jobs the law requires medical examinations (e.g. for driving HGVs) but pre-employment checks are not legally required for most jobs.

The Management of Health and Safety at Work Regulations 1999 say that you have to provide health and safety training for people when they start work, when their work or responsibilities change and/or periodically if their skills are not used regularly. You should bear in mind that people returning after illness may also need help readjusting to their job. Training must be provided during working hours and not at the expense of those working for you.

When someone starts work, they will need some form of induction training. They need to know the company's safety policy and the arrangements you have made to deal with health and safety matters. Think about the needs of young and inexperienced recruits who will need a more detailed induction and be aware of special needs, such as language differences. Many of the other regulations, for example, the Provision and Use of Work Equipment Regulations 1998 (PUWER) require that you train your staff.

You should also be aware of the Disability Discrimination Act 1995. It applies to employers with 15 or more employees. The Act makes it unlawful to discriminate against a disabled employee or job applicant by:

- treating him less favourably (without justification) than other employees or job applicants because of his disability; or
- not making reasonable adjustments (without justification).

We deal with this issue further in section 10 but we also suggest you either obtain a full copy of the Act from the Stationery Office (see the Appendices for details) or go onto the government's website at www.disability.gov.uk which provides some useful guidance.

Approaches to training

'Sitting by Nellie' (i.e. learning from an experienced person on the job) can be a very effective means of training. However, it does depend on how good 'Nellie' is at passing on her skills. Nellie may also need some training on how to train!

Other forms of training include outside trainers who can be brought on site, distance learning via videotape or computer-based learning. Consider whether there are any standards of competence for what you do, for example, NVQs.

Once training has been given it is a good idea to document who has given what training, when and to whom. Training should be repeated periodically where appropriate and adapted to take into account new or changed risks.

Health surveillance

More people die from work-related diseases than from workplace accidents. The Management of Health and Safety at Work Regulations 1999 and the Control of Substances Hazardous to Health Regulations 2002 (COSHH) require health surveillance in special cases. Some jobs that require surveillance include commercial diving, work with asbestos insulation and work with chemicals. Special conditions also apply to some people, such as pregnant women, whose jobs may expose them to lead or ionising radiation.

Health surveillance means having a system to assess early signs of ill-health caused by substances and other hazards at work. It includes keeping records, organising medical examinations and testing blood and urine samples. If there are known health risks from the work you undertake, or you are in doubt, take specialist advice.

Personal protective equipment

Even when risks have been assessed, and controlled and safe systems of work have been applied, some risks will still remain. These could include a risk of injury to the head, for example, from falling materials; to the lungs, from breathing in contaminated air or asbestos fibres; and to the eyes, from splashes of corrosive liquid. In these circumstances, personal protective equipment (PPE) is required to reduce the risk of injury.

PPE should only be used as a last resort if there is no other way of removing the risk of injury. If PPE is still needed, it must be hygienic and otherwise free of risk to health, and provided free by the employer.

Legal requirements

The Personal Protective Equipment At Work Regulations 1992 (PPE) lay down the main requirements. However, there are special regulations which cover lead, asbestos, hazardous substances, noise and radiation. The Construction (Head Protection) Regulations 1989 also apply. (See chapter 3, section 14 on construction for the main provisions.)

Risk assessment

If you run a small corner shop or a lawyer's practice, you may not need to consider protective clothing in any detail. However, in areas of manufacturing, construction, warehousing and distribution, there will be a risk of injury to staff.

Look for hazards

Consider:

- Who is exposed and to what?
- For how long?
- To what extent?

Reduce the risk of injury

Factors such as comfort, choice of equipment involved, ease of movement, ease of putting the item on and removing it, the effects of high temperatures and maintenance of parts are significant factors when considering the use of PPE as a means of protecting workers from hazards.

You should always take account of the health of the person who must wear the PPE. A high level of supervision and control is necessary to ensure constant use of the equipment. The employer should organise, where appropriate and at suitable intervals, demonstrations in the wearing of the PPE. Specific areas which can be protected include the following:

- **Eyes:** there may be hazards from chemical splashes, dust, gas and vapour, bright lights or radiation. Consider whether spectacles, goggles, face screens and/or helmets are necessary. Ensure that eye protection is the appropriate protection for the task. Ensure that a good quality product is bought and that the equipment suits the wearer in terms of size, fit, comfort and weight. Make sure that if more than one item of PPE is being worn, they can be used together. For example, a respirator may not give proper protection if the user is wearing safety glasses.

- **Head and neck:** beware of impact from falling or flying objects. There may also be a risk of hair entanglement or chemical drips and risks from climate or temperature (e.g. burns or frostbite). Consider helmets, hairnets, sou'westers and cape hoods. Neck protection may also be necessary, for example, during welding.

- **Ears:** refer to chapter 3, section 9 on noise. Consider earplugs or ear defenders. Only specially designed ear defenders should be fitted over safety helmets.

- **Hands and arms:** there may be a risk of cuts, punctures, chemicals, extremes of temperature, skin irritation or vibration to name a few. Consider gloves, gauntlets, mitts, wrist cuffs or armlets. Products are continually developing and you may wish to check with manufacturers on a regular basis as to whether you are using the most appropriate protection currently available. In a civil claim, where lack of PPE is alleged, you will be judged according to the types of equipment available at the time of the accident. You will also need to weigh up the risks from overheating as a result of wearing hand and arm protection (i.e. this may cause drowsiness) against the risks which are outlined above.

- **Feet and legs:** there may be a risk of slips, cuts and punctures, falling objects and chemical splashes. Consider shoes with steel toecaps, gaiters, leggings and safety boots. You will need to consider the appropriate sole pattern for the particular floor surface in your workplace.

- **Lungs:** there may be a risk from dusts, gasses and vapours. Various respirators and masks are available and specialist advice should be sought as to the most appropriate. Also consider the risks from overheating. All equipment should be suitable for its purpose and meet the necessary standards.

- **Whole body:** there may be a risk of extreme temperature, chemical splashes or clothing becoming entangled. The choice of materials for full body protection will depend upon the job in hand. Also consider other protection, for example, safety harnesses or life jackets.

As an accident prevention strategy, the use of protective equipment relies heavily on the worker wearing the item of PPE during the whole period of time they are being exposed to the hazard. Therefore, once PPE has been provided to your employees, you should ensure that it is being worn correctly. Safety signs can be a useful reminder (see below).

Signs

Where a risk has been identified and it cannot be eliminated, it may be appropriate to fit a sign in an appropriate place. The Health and Safety (Safety Signs and Signals) Regulations 1996 encourage standardisation of safety signs throughout the member states of the European Union so that signs, wherever they are seen, have the same meaning. The sign should be maintained in good condition and kept up to date. Some signs are not immediately self-explanatory and training or instruction should be given. The law requires an employer to:

- use road traffic signs within the workplace to regulate road traffic where necessary;

- maintain safety signs which are provided by him;

- explain unfamiliar signs to employees.

Examples of the most commonly used signs appear below:

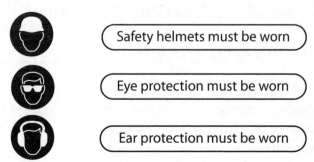

	General danger
	Industrial vehicles
	Corrosive material
	Flammable material
	Explosive material
	Toxic material
	No access to pedestrians
	Marking for dangerous locations
	No smoking
	Fire safety signs

Special needs/The Disability Discrimination Act

Protection of young persons

- Every employer shall ensure that young persons employed by him are protected at work from any risks that are a consequence of their lack of experience or absence of awareness due to the fact that they have not fully matured.

- No employer shall employ a young person for work:

  - which is beyond his physical or psychological capacity;

  - involving harmful exposure to toxic or carcinogenic agents;

  - involving harmful exposure to radiation;

  - where there is a risk to health from extreme cold, heat, noise or vibration.

The Disability Discrimination Act 1995

The Disability Discrimination Act (DDA) aims to end the discrimination which many disabled people face. This Act gives disabled people rights in the following areas:

- Employment

- Access to goods, facilities and services

- Buying or renting land or property (regarding access primarily)

The Act defines a disabled person as someone with a 'physical or mental impairment which has a substantial and long-term adverse effect on his ability to carry out normal day-to-day activities'.

You may need more information on who is likely to be covered by this Act. Statutory guidance exists which gives more background information on matters to be taken into account in determining questions relating to the definition of disability. A print copy of the guidance can be purchased from the Stationery Office (see the Appendices for their contact details).

The Disability Discrimination Act – employment provisions

- The employment provisions apply to employers with 15 or more employees. As stated the provisions, including those that require employers to consider making changes to the physical features of the premises that they occupy, have been in force since December 1996.

- There are two ways in which an employer may unlawfully discriminate against a disabled employee or job applicant:

  - by treating him less favourably (without justification) than other employees or job applicants because of his disability; or

  - by not making reasonable adjustments (without justification).

A code of practice – 'Elimination of discrimination in the field of employment against disabled persons or persons who have had a disability' – describes and gives general guidance on the main employment provisions of the Act. A print copy of this code of practice can be purchased from the Stationery Office.

The Disability Discrimination Act – access to goods and services

- Since October 1999, companies that provide services have had to consider making reasonable adjustments to the way they deliver their services so that disabled people can use them.

- Since October 2004, service providers have had to consider making permanent physical adjustments to their premises if it would otherwise be unreasonably difficult for disabled people to use their services.

Under the Workplace (Health, Safety and Welfare) Regulations 1992, recent amendments state that adequate seating must be provided in rest areas for the number of disabled persons at work. In addition, all parts of a workplace used or occupied directly by disabled persons at work must be organised to take into account their needs.

The Disability Rights Commission and the Disability Unit at the Department for Work and Pensions (see the Appendices for their contact details), through their helpline and website, provide information and advice both to disabled people on their rights and to service providers on their duties under the DDA.

New or expectant mothers

In many workplaces there are hazards which may affect the health and safety of new and expectant mothers, and that of their children. There are specific laws which require employers to protect the health and safety of these employees.

Legal requirements

The Management of Health and Safety at Work Regulations 1999 apply, as does the Sex Discrimination Act 1975. If an employer fails to protect the health and safety of their pregnant workers, it is automatically considered to be sex discrimination.

Risk assessment

The employer is required to conduct a risk assessment for his employees which should also include any specific risks to females of childbearing age who **could** become pregnant and any risks to new and expectant mothers. New mothers include those who have given birth within the previous six months or who are breast-feeding.

Look for hazards

Your employees must inform you that they are pregnant or breast-feeding. Until you have received written notification, you are not obliged to take any action other than those resulting from the risk assessment for all employees, mentioned above.

Once you have received notification, consider whether the employee may be at risk from different physical, biological and chemical agents, working conditions and processes.

Some of the more common risks may be as follows:

- The lifting or carrying of heavy loads.

- Standing or sitting for long periods of time.

- Exposure to infectious diseases, lead, radioactive material and smoke in the workplace.

- Work-related stress.

- Workstations and posture, as well as long working hours.

- Excessive noise.

Consider whether it is possible to remove or reduce the risk. For example, can the working conditions or hours of work be adjusted, or can the employee be given suitable alternative work on the same terms and conditions?

As an employer you will need to regularly monitor and review any assessment made to take into account the possible risks which may occur at the different stages of pregnancy. If it is not possible to reduce or eliminate the risk, then an employer must suspend an employee from work on paid leave for as long as is necessary to protect the health and safety of the employee and/or her child.

Working time

Working time is the time spent during your working hours which includes rest breaks. It is now controlled by legislation.

The Working Time Regulations 1998 impose obligations on employers in relation to the working time of workers over the minimum school-leaving age, including the provision of rest breaks and night work restrictions. The HSE and your local authority are the enforcing authorities.

The regulations apply to most (but not all) workers. For example, training doctors are excluded.

The regulations provide that:

- workers must not work for more than an average of 48 hours per seven-day week, calculated over a 17-week period;

- travelling to work is not included, unless travel is part of the job;

- workers can agree to work more than a 48-hour week but the agreement must be:

  - in writing;

  - in relation to a specified period or apply indefinitely;

  - terminable by the worker by seven days' notice given to the employer;

- the normal working hours of night workers must not exceed an average of eight hours in a 24-hour period. (There are, however, exceptions, e.g. surveillance or security workers.)

The first criminal prosecution under the Working Time Regulations took place in 2002 against a newsagents, for failing to limit an employee's hours to an average of 48 per week. The employee involved had worked up to 97 hours per week and was awarded £1,200 compensation. The company was fined £5,000 and ordered to pay £2,150 costs. These prosecutions will no doubt become more common.

See chapter 3, section 15 for information on tachographs and the legal restrictions for driving times and breaks.

CHAPTER 3

Specific problems in the workplace
and how to deal with them

Slipping and tripping

Slipping and tripping injuries are the single most common cause of injuries at work, costing employers over £300 million a year in lost production and related costs. The increasingly litigious nature of society is escalating slip and trip claims to an astronomical level. Solicitors are continually dealing with claims following slips on wet leaves in poorly lit car parks, on ice from leaking gutters and from spillages on the shop floor. These are all accidents which can easily be avoided. In a recent claim, a customer at a hairdressers succeeded in recovering damages after she slipped on an icy entrance step. The owner was found to have been negligent in not putting down salt, as the icy weather had been forecast.

Legal requirements

The Workplace (Health, Safety and Welfare) Regulations 1992 and the Management of Health and Safety at Work Regulations 1999 require employers to provide a safe workplace where floors, corridors and stairs are free from obstruction and debris. Outdoor routes must also be kept safe, particularly during wet and icy conditions. Precautions should also be put in place to prevent either people or materials falling. The Occupier's Liability Act 1957 also requires employers as occupiers to take reasonable care that anyone invited or permitted to be on their premises is reasonably safe.

Risk assessment

Look for hazards

Some of the most common types of accident in the working environment are caused by the following:

- Trailing cables
- Cardboard boxes causing obstruction
- Badly lit stairways and corridors
- Damaged floor coverings
- Spilt drinks

Reduce the risk of injury

- Employees should be told specifically that they must let the office manager or health and safety officer know of any of the above hazards as soon as possible.
- Floors should be in good condition and free from any obstructions.
- You may need to use separate routes for pedestrians and vehicles.
- Repair any uneven surfaces, holes or broken boards. Inspect them regularly and keep records.
- All areas including car parks and outside routes should be kept clean and records kept of inspection, cleaning and maintenance procedures.
- When floors are cleaned, ensure a slippery floor sign is clearly displayed. Even if you employ cleaning contractors, you will still have an obligation to your own staff.

- Ensure that employees are wearing the correct footwear if special footwear is necessary. If this is the case, you must pay for it.

- Ensure that there is an adequate level of lighting, including outside lights. Fittings and levels of lighting should be assessed regularly.

- Ensure that there are handrails on the stairs and ramps where necessary.

- Ensure that doors are safe. For example, consider fitting vision panels on swing doors.

- Clean up spillages immediately. Not only should spillages be cleaned promptly, it is vital that other employees are alerted to the fact that there is an area of slippery floor.

- If in-house repairs are insufficient consider contacting an expert, for example, the floor manufacturer, a specialist contractor or other health and safety expert.

- Ensure that the stair coverings and treads are in good repair.

- Think about marking steps and kerbs with black and yellow diagonal stripe tape.

- Ensure that there are arrangements in place to clear ice and snow, for example, putting down salt in cold weather.

- Specific guidance is provided for areas of particular risk (e.g. where the floor may become wet from condensation) from the British Standards Institution. They also provide guidance on lighting, stair/step measurements and managing risks generally (see the Appendices for their contact details).

Manual handling

A third of accidents reported at work are associated with manual handling. Manual handling covers a wide range of activities from pushing a wheelie bin to working at a checkout. It is defined as 'supporting or transporting loads by bodily force'. Many manual handling injuries build up over a period of time rather than being caused by a single handling accident.

Legal requirements

The employer's legal duties to prevent manual handling accidents are primarily found in the Manual Handling Operations Regulations 1992. They include the following:

- Employers must take all appropriate measures to avoid the need for any manual handling.

- Risks to employees must be assessed and where manual handling is unavoidable appropriate measures must be taken to minimise the risk of injury, particularly back injury.

- Planning and control. Managers and supervisors must devise and apply safe systems of working practice. They must also ensure that the right equipment and environment is provided. Load details, for example, the weight and which way up the load should be carried, should be given.

  This duty is quite onerous. Employers must consider maintenance work including emergency maintenance. In a recent case, a worker was carrying out necessary maintenance work on a conveyor system and had to remove a roller. The worker thought the roller would be hollow but it was solid and weighed around 20 kilograms. As he removed the last bolt, it trapped his hand causing an injury. As no assessment had been undertaken and no information had been provided to the worker, the employer was liable.

- Staff must be trained in the appropriate correct methods of handling and be shown how to recognise dangerous practices.

- Employees must be monitored for injuries and immediate remedial procedures must be put in place to reduce any new risks as they are identified.

- Employees must co-operate with employers and use the equipment and systems of work provided.

Risk assessment

Look for hazards

In undertaking the risk assessment, various aspects need to be looked at:

The task

Does it involve:

- the support and movement of a load?
- frequent or prolonged effort?
- twisting or stooping?
- insufficient rest or recovery?

The load

Is the load:

- heavy, sharp, hot or damaged?
- bulky, unwieldy or difficult to grasp?
- unstable or the contents likely to shift?

The working environment

Are there:

- extremes of temperature or humidity?
- space constraints preventing good posture?
- uneven, unstable or slippery floors?

The individual employee's capability

Does the job:

- require unusual strength or height and is the employee physically suitable to carry out the operation?

- require special clothing or footwear?

- create a hazard to those with health problems?

- require special knowledge or training?

All employees should be trained in safe lifting techniques. A booklet, entitled *Getting to Grips with Manual Handling*, deals with safe lifting and can be obtained from the Health and Safety Executive (HSE). The guidelines published in connection with the Manual Handling Operations Regulations 1992 provide guidance on the maximum lifting weights. As the HSE points out, there is no such thing as completely 'safe' manual handling but working within the above guidelines will reduce the risk and the need for a more detailed assessment. See the diagrams on pages 53–55 which show how men and women can lift safely.

Please note that the weights in the diagrams do assume that the load is readily grasped with both hands and that the operation takes place in reasonable working conditions with the lifter in a stable body position.

Reduce the risk of injury to the lowest level practicable

This means reducing the risk until the cost of any further precautions in time, trouble or money would be far too great in proportion to the benefits. This may include considering the following:

- **Task improvements:** this will include improving the layout of the task, the work routine and the work method, as well as reducing repetitive movements.

- **Changes to the load:** the load may be lighter, smaller, easier to manage, easier to grip, more stable and less damaging to hold.

- **Working environment:** consider space constraints, the floor and other aspects of the working environment.

- **Individual selection:** consider age, gender, health, fitness and the strength of an individual.

- **Mechanical assistance:** this would include the use of hand tools and mechanical lifting equipment.

- **Protective equipment:** for example, for hands and feet.

- **Training:** see the diagram below for the correct lifting technique. Leaflets including this information can be given to employees. Safe lifting needs to be planned. Employees must be trained to consider the following:

  - What they are lifting.

  - Its weight.

  - Its centre of gravity.

  - How to attach a load to the lifting machinery.

  - Who is in control of the lift?

  - The safe limits of the equipment.

- **Reporting:** encourage the reporting of aches and pains.

General risk assessment guidelines

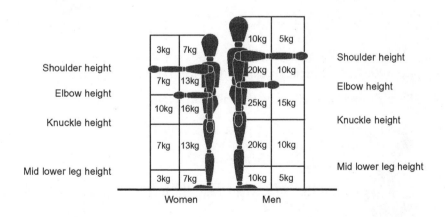

- Each box in the diagram above shows guideline weights for lifting and lowering.

- Observe the activity and compare it to the diagram. If the lifter's hands enter more than one box during the operation, use the smallest weight. Use an in-between weight if the hands are close to a boundary between boxes. If the operation must take place with the hands beyond the boxes, make a more detailed assessment.

- The weights assume that the load is readily grasped with both hands.

- The operation takes place in reasonable working conditions with the lifter in a stable body position.

- Any operation involving more than twice the guideline weights should be rigorously assessed even for very fit, well-trained individuals working under favourable conditions.

- There is no such thing as a completely 'safe' manual handling operation but working within the guidelines will cut the risk and reduce the need for a more detailed assessment.

Lifting techniques

- **Place the feet apart** giving a balanced and stable base for lifting with the leading leg as forward as is comfortable.

- **Adopt a good posture.**

- Bend the knees.

- Keep hands level with waist.

- Lean forward a little over the load.

- Keep the back straight (tucking in the chin helps).

- Keep shoulders level facing the same direction as the hips.

- **Get a firm grip.**

- Try to keep arms within the boundary formed by the legs.

- **Don t jerk.**

- Carry out the lifting movement smoothly, keeping control of the load.

- **Move the feet.**

- Don't twist the trunk when turning to the side.

- **Keep the load close** to the trunk.

- Keep the heaviest side of the load next to the trunk.

- If a close approach to the load is not possible, try sliding the load towards you before attempting to lift it.

- **Put the load down, then adjust.**

- If precise positioning of the load is necessary, put it down first, then slide it into the desired position.

Upper limb disorders

The term 'upper limb disorders' covers a whole range of problems affecting the neck down to the fingers, from tennis elbow, carpal tunnel syndrome, tenosynovitis to generalised repetitive strain injury. They include such unusually named conditions as vibration white finger, golfer's elbow and gamekeeper's thumb.

There have been many cases concerning the extent to which various upper limb disorders are attributable to problems in the workplace. It is generally accepted that many may be caused, or at least made worse, by certain types of work. Much depends on the working conditions, as well as an individual's medical history and susceptibility. Upper limb disorders can give rise to severely disabling symptoms, which can seem relatively trivial at first, but should be taken seriously.

In a recent case, a solicitor's firm was found to have been negligent and the claimant, a legal secretary, received over £85,000 in damages when she suffered a repetitive strain injury to her hands and wrists. The judge found that there were inadequacies with her workstation but that there were other contributory factors including tight deadlines, a heavy workload, poor posture, absence of proper breaks and no change in routine. The claimant was unable to continue with typing work due to her condition.

Legal requirements

The employer is required to assess risks by the Management of Health and Safety at Work Regulations 1999. There are no specific regulations but much guidance has been provided by the HSE and others. You should also refer to the next section on VDUs as there is some overlap between these two areas.

Risk assessment

Look for hazards

It is important to look at the duration of routine tasks requiring dexterity or arm movements, the number of repetitions, the force used and the posture adopted by the employee. For example, any employee carrying out routine copy typing all day may well be at risk if action is not taken.

Production line work is often very repetitive, with limited opportunities for breaks unless these are specifically organised. Look for work particularly involving hammering, gripping, twisting, reaching and squeezing. Are there any actual cases of upper limb disorders among the workforce or complaints of problems in upper limbs?

In the case of office workers, see the useful diagrams on pages 60 and 61 concerning posture and lighting and check whether each workstation complies with the recommendations.

Reduce the risk of injury

Once it has been decided that there is potentially a risk, the following action may be taken:

- Altering the office workstations to comply with the HSE's recommendations.
- Varying the work and alternating different types of work.

- Job rotation.

- Organising breaks and ensuring these are taken.

- Assessing and, if necessary, redesigning workstations.

- Checking each employee's individual posture and seeing how awkward positions can be avoided, for example, by providing steps or more space.

- Considering whether part of the process can be automated further.

- Providing better tools.

- Ensuring that the necessary retraining is given.

- Reducing time pressures if the workforce agrees, for example, by removing piecework/bonus schemes.

- Allowing self-pacing by workers if possible, rather than speed being dictated by machines.

- Ensuring better lighting. (This may avoid employees adopting awkward positions in order to see better.)

- Providing a warm environment if possible, or at least adequate protective clothing.

- Providing training in skills, posture, and warning symptoms for those at risk.

- Providing medical surveillance for those at risk and encouraging the reporting of symptoms.

Hand-arm vibration

If someone is exposed to too much vibration, for example, through using chain saws, hand-held power tools or road drills, they may develop hand-arm vibration syndrome. This is a group of symptoms – the best known being vibration white finger. There may be loss of sensation in the fingers, reduced dexterity, pain and increased sensitivity to cold. Continued use of power tools after symptoms start will mean that the symptoms will progressively worsen and may become irreversible. There is a risk of problems if, for example, the use of tools causes tingling or numbness, either during use or immediately after.

Many of the general comments relating to upper limb disorders apply here. There are some further specific considerations. Action to take also includes the following:

- Ensuring that tools are purchased with vibration control built in.

- Ensuring that employees take regular breaks.

- If there are existing problems, then cutting down the use of power tools as far as possible and looking at whether the job can be done another way.

- Ensuring that the tools are well maintained, as this should minimise the vibration produced.

- Keeping employees warm as far as possible, or at least providing gloves.

- Training employees to warm up and exercise their fingers in order to improve blood flow.

- Ensuring employees are trained to recognise symptoms and understand the importance of reporting these to avoid more serious problems developing.

- Creating a system of health surveillance may be appropriate.

VDUs

Visual display units or display screen equipment are commonplace in offices, normally as computer screens. Health risks posed by VDUs include the following:

- Postural, musculo-skeletal problems including upper limb disorders.

- Eye strain/discomfort.

- Headaches.

- Fatigue and stress.

- Flickering screens may affect some epileptics.

- In unusual circumstances, skin problems, possibly due to a combination of dry air, electrostatic charge and an individual's predisposition to such problems.

Controversy continues regarding the association between VDU usage and pregnancy or birth defects, but the official view of the HSE in the UK is that a risk does not exist. The Health and Safety (Display Screen Equipment) Regulations 1992 (DSE) require that pregnant women should be able to discuss any concerns with a knowledgeable person.

See also the previous section on upper limb disorders as many of the comments there also apply to VDUs.

Legal requirements

The Health and Safety (Display Screen Equipment) Regulations 1992 (DSE) require that employers do the following:

- Carry out risk assessments relating to relevant employee workstations. Relevant employees, generally, are those who habitually use VDUs as a significant part of their normal work.

- Ensure that workstations meet minimum requirements as set out in the regulations.

- Ensure that there are regular breaks in the work/changes of activities.

- On request, arrange and pay for eyesight tests and provide glasses if special ones are required.

- Provide health and safety training and information to employees.

The Workplace (Health, Safety and Welfare) Regulations 1992 also apply.

Risk assessment

Look for hazards

Careful assessment of the working conditions is needed and quite a high level of detailed consideration required. Certainly in an office environment there is considerable overlap with any assessment relating to the risk of upper limb disorders.

Seating and posture for typical office tasks

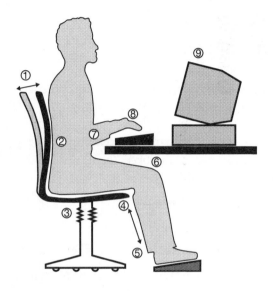

1. Seat back adjustability

2. Good lumbar support

3. Seat height adjustability from the seated position

4. No pressure on underside of thighs, knees and backs

5. Foot support for smaller users

6. Space to enable and encourage postural change

7. Forearms approximately horizontal

8. Minimal extension, flexion or deviation of wrists

9. Screen height and angle should allow comfortable head position

Lighting for typical office tasks

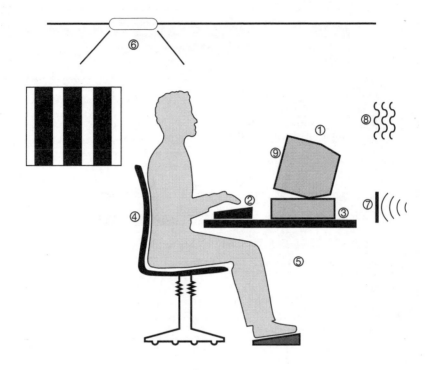

1. Screen: Readable and stable image, adjustable, glare free

2. Keyboard: Usable, adjustable, key tops legible

3. Work surface: Should allow flexible arrangement, be spacious and glare free. Document holder used if appropriate

4. Work chair: Appropriate adjustability plus foot rest

5. Leg room and clearances to facilitate postural change

6. Lighting: Provision of adequate contrast, no direct or indirect glare or reflections

7. Distracting noise minimised

8. No excessive heat, adequate humidity

9. Software appropriate to the task and adapted to user capabilities. Should provide feedback on system status

Example display screen equipment assessment record

White Young Green	**WORKSTATION ASSESSMENT**	Assessment No: **WYG. 251**

Department	Accounts
Location	Executive Park, Avalon Way, Anstey.
User Name	**Joanna Bloggs**
Job Description	Sales Ledger Clerk

Computer System	Mini Tower
Screen Model	17" CRT
Keyboard	Standard
Other Input Device/s Used (e.g. Mouse)	Wheel Mouse

Points to be Covered	Y / N	Assessor's Comments	Date to be Actioned
Is the VDU used for more than 2 hours each working day?	Y	7.5 hours	
Display Screen			
(a) Is the image on your screen clear & easy to read?	Y		
(b) Is the screen positioned at the correct (comfortable) height and distance?	N	Monitor raised on pile of old magazines. Recommend replaced with monitor stand. Eye line should be level with top of screen.	
(c) Can the screen be swivelled & tilted?	Y		
(d) Is the image on the screen free from flicker?	Y		
(e) Are there any distracting reflections or glare?	N		
(f) Is the screen damaged?	N		
(g) Is the screen clean and does the user have cleaning materials?	Y		
Keyboard			
(a) Is the keyboard separate from the screen?	Y		
(b) Is the keyboard tiltable?	Y		
(c) Is there sufficient space in front of the keyboard for hands / arms?	Y	Recommend move keyboard closer to prevent overstretching of arms.	
(d) Is the keyboard undamaged and are the keys clear?	Y		
Other Input Devices (e.g. Mouse)			
(a) Is your mouse within easy reach so that you do not have to stretch your arm to use it?	Y		

Example display screen equipment assessment record (continued)

White Young Green	**WORKSTATION ASSESSMENT**	Assessment No: **WYG. 251**

Points to be Covered	Y / N	Assessor's Comments	Date to be Actioned
Work Chair			
(a) Is the chair adjustable in height?	Y	Recommend height of chair raised so that a better postural position is achieved to reduce potential strain on the arms & shoulders.	
(b) Is the backrest adjustable in height and tiltable	Y		
Work Surface & Desk			
(a) Is there sufficient space to carry out required tasks & to allow for changes in posture?	Y	If desk pedestal were positioned under desk would limit space for changes in posture considerably.	
(b) Is the workstation arranged to avoid sitting in a twisted position?	Y		
(c) Is the desk surface matt / non-reflectant?	Y		
(d) Is the desk undamaged, stable and the right height?	Y		
Accessories			
(a) Do you have a document holder? (If NO, do you require one?)	N	Not Required	
(b) Mouse mat provided?	Y		
(c) Footrest provided? (If NO, do you require one?)	N	Not Required	
(d) Wristrests provided if required?	Y		
Work Arrangements			
(a) Do you vary screen based and non-screen based tasks?	Y		
(b) Have you been trained in the use of the relevant software?	Y		
(c) Have you been trained in adjustment of furniture, good posture and the DSE Regulations?		During Assessment.	
(d) Are you aware of how to report any problems and who to?	Y		
(e) Do you use a Laptop Computer?	N		
(f) If YES to (e) above, for how long each day?			
Health			
(a) Aware of need to report health problems immediately and who to?	Y	Line Manager	
(b) Aware of eye-sight testing procedure?	Y		

Example display screen equipment assessment record (continued)

WORKSTATION ASSESSMENT	Assessment No: WYG. 251

Department	Accounts
Location	Executive Park, Avalon Way, Anstey.
User Name	**Joanna Bloggs**
Job Description	Sales Ledger Clerk

Computer System	Mini Tower
Screen Model	17" CRT
Keyboard	Standard
Other Input Device/s Used (e.g. Mouse)	Wheel Mouse

Remedial Measures Necessary	Confirmation of Action Taken	Date
Monitor raised on a pile of old magazines. Recommend replace with a monitor stand so that eye line is level with the top of the visible screen.		
Recommend move keyboard closer when it is being used for continuous periods of typing to prevent overstretching of arms.		
It is recommended that the height of the users chair is raised so that a better postural position is achieved to reduce potential strain on the arms, back & shoulders.		
If desk pedestal were positioned under desk this would restrict space for changes in posture considerably. Desk pedestal should not be positioned under desk but alternative, narrower pedestals or repositioning of rows of desks should be investigated so that the width of aisles between desks is not reduced to an extent where access and egress, especially for emergency evacuation, is restricted.		

Assessment Date	Signature of Manager	Signature of User

Reproduced by the kind permission of White Young Green

A useful basic form to assist with this risk assessment appears on pages 62–64. More detailed forms are available, for example, from the Royal Society for the Prevention of Accidents (RoSPA) – for their address, see the Appendices. Also, you may wish to ask each user to assess his workstation at the beginning of a review. This should show any area where urgent action is required.

Reduce the risk of injury

Action may include the following:

- Reviewing each employee's posture whilst at the workstation and adjusting the equipment accordingly.

- Scheduling regular breaks and ensuring that they are taken.

- Providing/ensuring variety in the work undertaken.

- Providing free eye tests and glasses if needed because of VDU work.

- Ensuring that the screens have adjustable brightness and contrast controls (compulsory since 1996).

- Ensuring that no bright lights are reflected in the screen and that the screen does not face bright lights or windows.

- When using a mouse, place the mouse close so it can be used with a relaxed arm and straight wrist.

- Ensure regular warnings are given to employees about the risks we have highlighted.

Revised guidelines, entitled *Working with VDUs*, were published by the HSE in June 2003. They can be obtained directly from the HSE.

Hazardous substances

Many substances are potentially hazardous to health, from asbestos to prawns, causing health problems from mesothelioma to prawn blower's asthma. This is a large topic and any employer working with chemicals and other potentially hazardous substances to any great extent should seek specialist advice, if not an expert.

Hazards may often be partly hidden, for example, asbestos pipe lagging in older buildings. If in doubt, an expert assessment should be made. There are cases where women have developed asbestos-related fatal conditions, such as mesothelioma, where there has been no apparent contact with asbestos in the workplace. However, on further investigation, it was proven that they had been exposed to asbestos when they had placed their husband's overalls in their washing machines.

At the end of this section, the question of smoking in the workplace and asbestos-related illness will be dealt with separately.

Legal requirements

The Control of Substances Hazardous to Health Regulations 2002 (COSHH) govern this area. Some substances, for example, lead, ionising radiation and explosives are covered by specific regulations and are not dealt with in detail here. See the regulations in the Appendices at the end of the book. Regulations covering suppliers of chemicals and the carriage of dangerous goods are also beyond the scope of this book.

Under the COSHH regulations, the employer needs to follow seven steps:

1. Assess the risks to health arising from hazardous substances in the workplace.

2. Decide what precautions are needed.

3. Prevent or adequately control exposure.

4. Ensure that control measures are used and maintained.

5. Monitor the exposure of employees to hazardous substances, if necessary.

6. Carry out appropriate health surveillance where necessary.

7. Ensure that employees are properly informed, trained and supervised.

Risk assessment

The health and safety policy should list the people responsible for COSHH risk assessments, the review period and the names of others who help to implement the action required.

For an example basic form for an initial COSHH risk assessment, see the following page.

Look for hazards

The main source of information on the qualities of any substance in the workplace is likely to be the supplier. Read the label and any data sheet. If a data sheet has not been supplied, the supplier should supply one on request. Then consider whether any of the possible hazards are likely to arise in the workplace, generally or in any particular circumstance. Then review the existing controls and consider whether anything further should be done.

A statement in a health and safety policy following a risk assessment relating to a standard office building may read as follows:

'In accordance with the COSHH regulations, an assessment has been carried out of the possible dangers of using various substances at work. It is concluded that the only dangerous substances are:

- **Toner:** used in photocopiers, printers and fax machines. In case of spillage, do not inhale the fine powder; in case of skin contact, wash with soap and water; in case of eye contact, treat as a foreign body.

- **Bleach and similar cleaning products:** stored in our cleaning cupboards. These can only be used by staff under the supervision of the health and safety officer or office manager.

- **Tippex and thinners:** it is considered that the likelihood of anyone suffering harmful effects from exposure is low and only likely to happen if the substances were to be used in some unusual way. To avoid any risk, please use these substances at work strictly in accordance with the manufacturer's instructions'.

COSHH Risk Assessment

Premises/work area: Date of assessment:

Assessment carried out by: Position:

Material	
Chemical component	
Quantity used	
Form of supply/use	
Information from suppliers: Available Adequate	
Internal date sheet prepared	
Main hazard present (gas, vapour, fume, liquid, dust, etc)	
Summary of hazard (including effect)	
Hazard identified	
Who is at risk?	
Existing controls	
Action required	

Review date:

Signed:

Then details would follow of the people responsible under the COSHH regulations for risk assessments and the implementation of any action, as well as details of the review period.

Other substances potentially occurring in an office environment may include ozone from photocopiers. The maintenance of the air conditioning also needs to be kept up to date.

Reduce the risk of injury

Following the risk assessment, action may be needed. Remember that exposure should be prevented if at all possible. If not possible, then it must be controlled. Action may include:

- Ensuring that clear labelling is present on all containers. Hazardous substances must be marked with a black and orange label.

- Ensuring that clear signs are present warning of hazards. See the examples in chapter 2, section 9.

- Providing suitable containers for all substances, including waste materials that may give off fumes for example, and ensuring that they are covered when appropriate.

- Considering a substitute for a substance, for example, if a chemical is flammable, looking for one with a higher flash point; if carcinogenic, looking for one that is not known to be carcinogenic.

- Reducing the amount used.

- Reducing the amount of contact an employee has with the substance.

- Providing good washing and changing facilities.

- Providing protective clothing, for example, to prevent contact with the skin or the inhalation of fumes.

- Not allowing smoking/drinking/eating in a chemical handling area.

- Reminding employees not to put pens and pencils in their mouth; this may transfer chemicals to their mouth in the process.

- Using a pump or hand-operated siphon to move chemicals.

- Controlling access.

- Taking fire precautions.

- Ensuring that substances are stored in accordance with the manufacturer's recommendations. If they are flammable, keeping them in secure storage with regulated temperature and ventilation, away from sources of ignition.

- Storing gas cylinders with valves uppermost, away from the main workplace, on the outside of the building if possible, above ground level. Also moving them in a way that minimises damage; using correct hoses, etc for the cylinder; turning off cylinder valves at the end of the day; changing cylinders in a well ventilated place; never using flame to test for leaks but using detergent/water solution instead; training welders in correct working procedures; fitting non-return valves and flame arresters; and ensuring sufficient ventilation.

- Ensuring regular cleaning. If dust is a problem, using a 'dust free' method, for example, a vacuum system with an efficient filter.

- If the plant atmosphere could be flammable, avoiding dust build up, using exhaust ventilation, controlling sources of heat and reducing any sparking.

- Improving ventilation.

- Informing and training all employees and keeping this under review.

Once action has been decided on and implemented, it must be maintained and kept under regular review at five-yearly intervals, if not more frequently.

Smoking

This can be an emotive issue. It is particularly important to consult the workforce before implementing any policy changes and to allow time for the adjustment to any changes. Under the Workplace (Health, Safety and Welfare) Regulations 1992, employers should protect non-smoking employees from the nuisance of tobacco fumes in communal rest areas. There are also strong recommendations from the government generally, and the HSE in particular, that non-smoking employees should not routinely be exposed to tobacco smoke in the workplace. There is now plentiful evidence that passive smoking is harmful to your health. Tobacco fumes can also exacerbate other health problems, for example, asthma.

Many employers operate a no smoking policy, due to the concern regarding the possible effects of passive smoking, as well as from pressure from non-smoking employees who find a smoky atmosphere unpleasant.

Depending on the size of the workplace, measures may include designating a separate smoking area, improving ventilation or at least limiting the time an employee is exposed to smoke.

Employees often end up smoking outside the workplace. If so, ensure that they are not causing a nuisance outside and, for example, where practical, provide a metal bin for cigarette ends.

Asbestos

Much has already been done to control the risk from asbestos through regulations prohibiting the use, supply and importation of asbestos and asbestos products. There are also regulations that control any work with, on or around asbestos, with many types of work requiring a licence. Specialist advice should be sought if your work involves coming into regular contact with asbestos.

Current regulations require that work with many asbestos products, particularly asbestos insulation, coating or insulating board, is carried out by contractors licensed by the HSE. A list of all contractors holding an HSE licence is available through the HSE.

For the most part, the Control of Asbestos at Work Regulations came into force in November 2002. However, because Regulation 4 creates onerous and potentially costly obligations, its implementation was delayed so as to allow property owners sufficient time to put into place the procedures for compliance. Regulation 4 therefore came into force on 21 May 2004.

Regulation 4 applies to all non-domestic premises and states that there is a duty on all those who have responsibility for the maintenance and/or repair of non-domestic premises to do all of the following:

- Investigate whether there are any materials in the premises likely to contain asbestos.
- Carry out a written risk assessment based on the investigation.

- Prepare and implement a risk management system, which includes the preparation and maintenance of a written plan identifying the location and condition of asbestos (if any).

- Provide anyone who is likely to work on or disturb such materials with information about their location and condition (including the emergency services).

This duty could extend to lenders, investors, developers and companies acquiring other businesses – all of whom should be vigilant to avoid inheriting asbestos liabilities. However, the regulations ask for a proportionate approach and only require a substantial survey where the risk warrants it. For example, in a small shop, a walk through inspection may be adequate. It is clear that the HSE will enforce the asbestos regulations strictly and will be looking to impose large fines.

Workers away from base

A substantial number of people work away from normal base operations. These include home and lone workers.

Home workers are those people employed to work at home for an employer and lone workers are those who are working away from the main work premises. Office cleaners, insurance salesmen, servers in coffee kiosks and home helps are a few examples.

This chapter also specifically deals with the special requirements of expectant mothers or those with infants. In addition, you should refer to the section on violence and consider whether other sections, particularly the sections on VDUs, stress and working time, also apply.

Legal requirements

Every business owner has a legal responsibility to protect the health and safety of their workforce. This includes staff who work on their own, either at the business premises or at some other location. Most of the regulations therefore apply to home or lone workers as well as to employees working at an employer's workplace.

Importantly, under the Management of Health and Safety at Work Regulations 1999, employers are also required to undertake a risk assessment of the work activities carried out by home workers.

The Occupier's Liability Acts of 1957 impose a duty on occupiers of the premises to take care of visitors in respect of dangers due to the state of the premises or things done or omitted to be done to it.

In addition, the Trade Union Reform and Employment Rights Act 1993 allows employees, including home workers, to stop working in the event of serious or imminent danger arising from the work that they are doing, without this affecting their employment rights.

Home workers, like other employees, have a duty to report all faults which may be a hazard to their own or other people's health and safety.

The Reporting of Injuries, Diseases and Dangerous Occurrence Regulations 1995 (RIDDOR) apply equally to home or lone workers (see chapter 4, section 1).

Risk assessment

Look for hazards

It may be necessary for employers to visit home workers to carry out a risk assessment. Home workers can help in identifying the hazards for their employers. In particular, consider the following:

- Do your employees have any medical condition that could compromise their health or safety?

- Could there be undue pressure caused by the work?

- Are you supplying specialist equipment for home workers?

- Is there any danger to other members of the family?

- Does any machinery need an electrical supply? If so, the business owner (not the manufacturer) is responsible for its safety. However, the safety of the electrical supply in the home (e.g. wall sockets) is the responsibility of the homeowner.

- Are any hazardous substances or chemicals being used?

- Does the work involve handling heavy loads?

- Special care must be given to expectant mothers and those with infants. The law requires that employers take account of the special circumstances of mothers-to-be and mothers who are breast-feeding or who have infants younger than six months. The employer should be aware that any risks affecting the mother would also include the child.

- Is there a danger of violence to those working alone? (See section 7 on violence.)

Decide who may be harmed and how

This may include home workers and members of their household, including children or visitors.

Reduce the risk of injury

- To avoid undue pressure, set reasonable limits on the amount of work that should be done in a period of time. Workers need to be trained. Also ensure that they are in regular contact wherever possible. Mobile phones or pagers could be encouraged and, if necessary, supplied. It may be useful to supply guidelines that may be followed in an emergency, listing contact names and useful telephone numbers.

- If equipment has been supplied, provide written instructions on how the equipment must be used and stored at the end of each day. Also:

  - ensure that the equipment is correct for the job;

  - provide the correct information and training;

  - check that the equipment is regularly maintained (and repair it when necessary);

  - ensure that the necessary personal protective equipment is provided (e.g. gloves when working with needles);

  - check that the plugs are not damaged and are correctly wired and maintained;

- repair electrical equipment which may cause harm or injury;

- check that there are no trailing wires and if there are any, tuck them out of the way, for example, under a desk or table to prevent accidents.

- If hazardous substances or chemicals are being used, these should be identified with a black and orange label on the container. You must also provide the home worker with a set of instructions on how to handle the substance.

Here is a checklist of things to be aware of when using substances, materials or chemicals that may be hazardous to health and safety (see also section 5 on hazardous substances):

1. Are they flammable, toxic or corrosive?

2. Are they stored safely, for example, could children reach them easily?

3. Does anyone suffer from skin rashes, irritation, dizzy spells and sickness or have headaches when coming into contact with the chemical?

4. Does anyone suffer from asthma?

Do note that employers are only responsible for the substances and materials they provide to their home workers.

- Recommend work break periods and inform the home worker that other members of the family, who are unskilled and inexperienced, must not become involved with the work.

- If the work involves lifting heavy loads, the necessary lifting aids should be provided (see section 2 on manual handling).

- It is important for workers using VDUs to adjust the workstation to a comfortable position and take regular breaks from work. VDUs need to be placed in a position where lighting will not cause reflection or a glare on the screen. VDU users can request an eye examination and eye test from their employer. (For more information, see section 3 on upper limb disorders and section 4 on VDUs.)

- Under the Health and Safety (First Aid) Regulations 1981, you must ensure that adequate first aid provision is supplied for home workers. The exact provisions depend on the nature of the work activity and the risks involved.

Other points to note

HSE inspectors have the right to visit home workers. They can also investigate and help settle complaints about working conditions that could affect the health, safety or welfare of all employees, including home or lone workers.

Violence at work

Violence can consist of any verbal abuse or physical threat upwards. It may be between employees or, more likely perhaps, come from members of the public, depending on the nature of the business.

Legal requirements

The employer is under the usual general duty under the Health and Safety at Work Act 1974 (the HSW Act) to ensure, so far as is reasonably practicable, that the health safety and welfare of his employees is protected at work. Where there is a risk to employees (and in some cases, persons not in his employment) of physical violence in particular, as with a jeweller's shop or a 24-hour service supermarket, an employer must be particularly aware of his duties under the Management of Health and Safety at Work Regulations 1999, particularly with regard to risk assessments.

Certain businesses may need a clearly defined policy on violence at work. Such a policy, although relatively short, should be incorporated as a subsection to the statement of the health and safety policy.

Risk assessment

Look for hazards

An obvious example of an area where there may be a risk of violence would be when employees have to meet customers on their own, particularly if this is away from the usual workplace. When looking at whether there is a risk, it is very important in this area for employers to speak to their employees. Individual employees may have suffered some incidents of verbal abuse for example, which they may not have reported. When everyone's experience is considered, there may be more of a problem than it first appeared. You also need to think about what might have happened in a given situation if the violence had escalated. Any problems encountered by other similar businesses should also be considered.

Also consider the risk to members of the public. In a recent case, Mr M was stabbed in the back by a bouncer employed by the defendant, a nightclub owner, and, as a result, was left a paraplegic. The stabbing took place following an altercation earlier in the evening at the nightclub. At the close of the evening the bouncer went home, got a knife and returned, whereupon he met up with Mr M. In deciding whether or not the employer was responsible for his employee's conduct, the court considered the closeness of the connection between the act of the employee and the duties he was engaged to perform. In this case, it was stressed that if the job required an employee to use violence, then an act of violence will be more likely to be found to be linked to the employment. The employer was therefore responsible for Mr M's injuries.

Reduce the risk of injury

Measures to avoid or reduce the risk include reviewing an employee's training. Are they trained to spot the risk? Can they tell quickly if a customer is becoming angry and have they been trained to deal with this? Do they know of any emergency procedures, for example, the use of a panic button? Is there a policy for dealing with complaints and difficult customers? Where are customers seen? Is it possible to ensure that two employees are present at all times? If not, are there surveillance cameras and is someone always within call?

Consider whether there is sufficient space between the employee and the customer. Screens may be advisable but in some circumstances these can exacerbate the problem by the atmosphere they can create.

Are any valuables locked away out of sight where possible? Are there time-delay operated safes?

Particular risks are posed by employees visiting customers outside the workplace. If this is unavoidable, consider whether two employees should go or at least deliberate the extent to which a customer should be 'vetted' beforehand. Do ensure that the employee leaves clear information about where he is going and for how long, and has a means of calling for help quickly if necessary, for example, via a mobile phone. Consider personal alarms for high-risk staff.

Do not overlook risks posed by travel, particularly by those travelling late at night. Do consider providing safe transport home and do ensure that all areas outside the workplace are well lit.

If an incident does occur, this should prompt a thorough review of the measures in place. Also ensure that any affected employees are well supported following such an incident. In serious cases, a victim support scheme may help. The local police station will know where the nearest scheme is.

Stress

Stress, a reaction to undue pressure, can lead to mental and physical ill-health. It is uncertain whether stress at work is actually increasing or just appears to be, but it is definitely the case that employees are more aware that they may have a claim against their employer as a result. Insurers are certainly encouraging their insured to immediately notify them of any claim where stress, bullying, harassment or discrimination is alleged.

With the benefit of hindsight, it seems many stress claims could have been avoided by improved personnel procedures and communications, and by creating a more open and helpful culture in the workplace.

Legal requirements

The Court of Appeal gave comprehensive guidance in this area in February 2002 in the case of Sutherland v Hatton. Their guidelines included some practical advice for the employer as follows:

- Is psychiatric harm to any particular employee reasonably foreseeable?

- An employer is expected to know things that he ought reasonably to have known about a particular employee, but an employer can assume that an employee can withstand the normal pressures of the job unless he knows that the employee has a particular problem or vulnerability.

- The same test applies to each employee, whatever their job.

- The employer should consider the nature and extent of the work done by the employee and signs from the employee of impending harm to health. Such signs must be plain enough for any reasonable employer to realise that he should do something about it.

- The employer is usually entitled to take what he is told by the employee at face value, unless he has good reason to think that what he is told is incorrect.

- The employer is only expected to take steps that are reasonable in the circumstances. The court will take into account the size, resources and demands on the employer's business.

- The employer is only expected to take steps that are likely to be effective.

- An employer who offers a confidential advice service, with referral to appropriate counselling or treatment services, is unlikely to be found liable to his employee.

- The employer is not liable for allowing a willing employee to continue in the job if the only reasonable and effective step would have been to dismiss or demote the employee.

To sum up, if it is possible to provide a confidential advice service with referral for counselling or treatment services, this would be the ideal. For example, providing the use of various stress telephone services (see the Appendices for the contact details for BUPA and the Centre for Stress Management who provide this facility). To be fair, this may well be beyond the scope of many small businesses and may not be required in any event. However, even the provision of such a service does not remove the need for an employer to consider the risk and take other effective action if it is indicated.

Ideally, the employer should be aiming to prevent stress from becoming a problem in the first place, but work-related stress is challenging to manage. Without always being able to determine the real source of stress, it is extremely difficult for employers to introduce stress management processes into the workplace. Careful risk assessments are therefore vital.

Risk assessment

Look for hazards

The usual risk assessment rules apply. Consider firstly whether there may be a problem. For example:

- Do your employees have poor sickness records?

- Are illnesses reported that may be stress-related (this covers a multitude of illnesses from a headache to irritable bowel syndrome and many others)?

- Is there a high staff turnover, poor time keeping and unhappiness among the employees or customers?

- Can any stress factors be identified? Talk to the staff.

- Is bullying and harassment sufficiently discouraged?

- Is there a culture of long hours, lack of communication and blame when things go wrong?

- Is there too much work for the number of staff? Look for pressures that could cause high and continuing levels of stress.

Then consider who may be harmed by any stress factors you may have found and what is reasonably practicable to do about it. Even if there are no obvious stress factors, it is worthwhile reviewing possible preventive measures.

Reduce the risk of injury

Preventive action may include:

- Encouraging an open and understanding culture.

- Ensuring good communications.

- Being understanding if employees admit to being under too much pressure.

- Ensuring that employees are properly trained for their work.

- Ensuring that employees have sufficient support.

- Allowing employees as much control as possible over their work and working conditions.

- Ensuring that employees are treated fairly.

- Preventing bullying and harassment by replacing line managers if necessary.

- Preventing bullying and harassment.

- Regularly monitoring the workforce for possible signs of stress.

- Providing a confidential advice service.

Action if there is a problem with workplace stress may include:

- Trying to address the sources of stress.

- Reviewing the preventive measures above and how effective they are for all employees. Also drawing up a further action plan if necessary.

- Discussing with the employee concerned and trying to involve him in decisions.

- Considering whether the employee should be moved or work fewer hours, or otherwise varying his work. If he is off sick, keeping in touch. Beware of the employment law implications of changing the terms of an employee's contract or dismissing him.

- If necessary, enabling and encouraging the employee to seek further help through their doctor or a counselling service.

Noise

Noise at work can cause damage to hearing and high levels can cause deafness. It can also cause tinnitus (irritating ringing in the ear), communication difficulties and stress.

It is estimated that two per cent of the working population in the UK are now employed in call centres. Here, acoustic shock/trauma – the result of sudden, specific loud noises coming over the telephone headset – is emerging as a new cause for claims. This can cause hearing loss, dizziness and headaches.

However, as with work-related stress, it is extremely difficult to quantify the risks to your employees, as people have different sensitivities to noise.

Exposure to noise may affect hearing in three ways:

- **Temporary threshold shift:** this is a short-term effect, is reversible and depends to some extent upon an individual's susceptibility.

- **Permanent threshold shift:** recovery will take place but not fully, and only after the end of the exposure.

- **Acoustic shock/trauma:** as mentioned above, this involves sudden aural damages resulting from short-term intense exposure or one single exposure, for example, fireworks or major explosions.

It is vital, therefore, for employers to monitor carefully and control the exposure of all employees to noise from work activities.

Legal requirements

The Noise at Work Regulations 1989 require that employers:

- make and regularly review noise assessments;
- keep records of assessments;
- generally reduce the risk of damage to the hearing of employees;
- reduce exposure, other than by use of ear protectors;
- provide ear protectors;
- make designated ear protection zones;
- use and maintain equipment to protect hearing;
- provide information, instruction and training for employees.

Employees have a duty to use the equipment and to report defects. Manufacturers and suppliers of equipment have a legal duty to provide information on the noise a piece of equipment produces. If such information is not clear, then manufacturers or suppliers should be asked to clarify the situation.

Daily personal noise exposure

The total amount of noise exposure during the whole working day is called 'the daily personal noise exposure'. Three 'action levels' are specified in the regulations:

First action level means a daily personal noise exposure of 85 dB(A). The employer must do the following:

1. Have the risk assessed by a competent person.
2. Tell the workers about the risks and precautions.
3. Make hearing protection readily available to those who want it.
4. Suggest workers take medical advice if they think their hearing is affected.

Second action level means a daily personal noise exposure of 90 dB(A). The employer must do the following:

1. Do all he can to reduce exposure as above, as well as providing hearing protection.
2. Mark zones where noise reaches the second peak action levels with recognised signs to restrict entry, for example, 'people must not enter these zones unless they are wearing hearing protection'.
3. Plastic foam and mineral fibre waxed plugs would be useful as hearing protectors.

The peak action level means a level of peak sound pressure of 200 pascals. Refer to the second action level as the same duties apply.

Risk assessment

Look for hazards

The regulations require that every employer should, when any of his employees are likely to be exposed to the first action level or above, or the peak action level or above, ensure that a competent person makes a noise assessment. Employers should look at how noisy equipment will actually be used on site. The person using the equipment ought to be able to talk to someone two metres away without having to shout to be understood. If they have to shout, the noise from the equipment is probably loud enough to damage their hearing, so action will be necessary.

However, the best way of assessing noise levels is employing someone with the necessary skill and experience to measure noise. See the Appendices for the contact details of the Association of Noise Consultants.

Reduce the risk of injury

- The regulations stress that if there is any likelihood of any damage, ear defenders or plugs ought to be provided.

- Ensure that any equipment provided to reduce noise exposure is actually used by employees.

- Choose the quietest equipment. It may prove useful to get some help from an independent consultant or the suppliers of the new machinery. In a recent case, a policewoman was awarded substantial damages when she developed tinnitus caused by excessive noise, generated by a receiver worn during surveillance work for the Serious and Organised Crime Squad. If the defendant had addressed the problem of noise exposure, it would have been comparatively simple to provide an earpiece with a limited output and an on-off switch.

- Reduce the time that employees are exposed. Hearing damage is cumulative and depends not just on the noise levels, but on how long people are exposed to them.

- Put noisy machines in separate rooms.

- Fit silencers to equipment, such as exhausts.

- Use sound insulation materials where possible.

- Provide hearing tests for those identified as being at risk.

- Employees are also under an obligation to use noise control equipment if provided, for example, use ear protectors particularly in any 'ear protection zones'. Once provided with ear protectors, employees must keep them in a sensible place and if there is any problem with them, they should draw their employer's attention to their difficulties.

Vibration

Vibration is often associated with noise (see the previous section). Vibration is difficult to measure but excessive exposure can lead to either 'hand-arm vibration syndrome' (HAVS) or 'whole body vibration' (WBV). The former is a painful condition affecting blood circulation, nerves, muscles and bones in the hands and arms. It is caused by excessive use of hand-held power tools and machinery such as chipping hammers, grinders and chain saws. The latter is a condition that mainly affects drivers of heavy goods vehicles, tractors and dumpers; and can also cause low back pain and spinal injuries. See also section 3 on upper limb disorders as much of the information found there also applies.

Legal requirements

Unlike noise, there are no specific recommendations for vibration but, where a hazard exists, the Provision and Use of Work Equipment Regulations 1998 (PUWER) and the Management of Health and Safety at Work Regulations 1999 apply.

Risk assessment

Look for hazards

- The hazards are similar to those highlighted in the section on upper limb disorders.

- Employers should consider vibration levels when deciding what equipment or machinery to purchase.

- Are there any cases of vibration-related disorders in your workplace?

Reduce the risk of injury

- Efforts should be made when carrying out a job to limit the use of high vibration tools. Consider whether the process can be automated further.

- Tools should be checked regularly to minimise vibration and, if necessary, anti-vibration monitors should be fitted.

- Vary the work by alternating the different types of work and introducing job rotation.

- Provide training in skills, posture and warning symptoms to those at risk.

- Depending on the nature of the job, gloves can be used, although it should be emphasised that there are no gloves that can adequately protect people from vibration.

- When working with such machinery, it is important to have regular breaks. Workers can then keep their hands warm by massaging their fingers.

- Medical surveillance should be provided for those at risk.

- Ultimately, it is a matter of common sense that overuse of certain types of machinery will cause vibration, heat tingling and numbness in the hands.

Electricity

The principal hazards associated with the use of electricity are the risk of electrocution, burns, fire and explosions.

Each year, about 1,000 accidents at work involving shock and burns are reported to the HSE and about 30 of these are fatal. Fires can be started by anything, from poor electrical installation to explosions caused by static electricity igniting flammable vapours or dusts.

It is therefore fundamental to any business that all electrical equipment, wiring installations, meters and wires connected to them are maintained to the highest standard.

Regular checks and inspections are necessary, but how often these checks/inspections should be carried out will depend on the nature of the electrical equipment.

Legal requirements

The legal position is governed by the Electricity at Work Regulations 1989. Further guidance is given in the Memorandum of Guidance accompanying the regulations. This book can only give a brief overview of what is required. Briefly, Regulation 16 requires that no person shall be engaged in any work activity where technical knowledge or experience is necessary, in order to prevent exposure to danger or injury.

The regulations apply to all workplaces and cover all electrical equipment ranging from battery-operated equipment to high voltage installations.

Risk assessment

Shock

Electric shock results from an electric current flowing through the body. The effect can be fatal if the heart rhythm is disturbed for long enough so that it prevents blood flowing to the brain. Emergency action is vital in such cases. Resuscitation must be commenced quickly, firstly ensuring that there is no risk that the first aider will also suffer an electric shock.

Consider displaying a copy of the HSE's poster Electric Shock Action which shows what to do in an emergency.

Employers must prevent access to electrical danger. In a recent case, a 15 year-old claimant climbed over an eight foot steel fence into the defendant's electrical substation compound, ignoring clear signs warning him of the deadly risk he ran by entering. Once inside, he climbed up between the cooling fins and received an electrical shock from the overhead bars. The defendant was still held to be negligent for not providing a fence which efficiently protected the claimant.

Principal protective measures against electric shock are as follows:

- Protection against direct contact, i.e. by providing proper insulation for those parts of the equipment liable to be electrically charged.

- Protection against indirect contact, i.e. by providing effective earthing for metallic enclosures which are liable to be charged with electricity if the basic insulation fails for any reason.

- Fuses and circuit breakers should be of a type and rating appropriate to the circuit.

- Reduced voltage systems.

- Ensure that mains switches are easily accessible and clearly identified so that fuses and other devices are correctly rated for the circuit they protect.

- There must be a switch or isolator near each fixed machine to cut off power in an emergency.

- Enough socket outlets must be provided and all plug sockets and fittings must be sufficiently robust.

- Anyone carrying out electrical work must be competent to do so safely. If this means employing independent outside contractors, then do so.

- Cables, bulbs and other equipment must be replaced regularly and kept in good condition at all times. Suspect or faulty equipment should be taken out of use and labelled 'do not use'.

Appliances

- Any electrical appliances should be unplugged before cleaning or making adjustments. It is important to carry out regular checks and repairs, ideally by a qualified electrician. Even normal mains voltage (240 volts AC) can cause fatalities.

- Explosions can be caused by electrical apparatus or static electricity igniting inflammable vapours or dusts. Employers should ensure that waterproof or explosion protected equipment is supplied.

- The risks are greatest when electricity is used in harsh conditions, for example, portable equipment used outdoors.

- Reduce the voltage. Lighting can run at 12 or 25 volts and portable tools at 110 from an isolating transformer.

- Contact with overhead electric lines accounts for half of the fatal electrical accidents each year. In circumstances where any equipment may come within nine metres of a power line, do not work under them without seeking advice from your electricity company. Electricity can flash from overhead power lines even if the plant machinery and equipment is not touching them.

Fire

Clearly any outbreak of fire will be hazardous and it is crucial that employers have a structured, well-organised policy for preventing fires occurring. See chapter 2, section 3 for more information. However, this will include such practical considerations as:

1. keeping the workplace tidy;

2. ensuring that emergency routes are free from obstruction;

3. designating competent persons who are responsible for ensuring that all plant machinery and equipment (that could potentially cause fire) is switched off every evening.

The local fire brigade, who are responsible for enforcing fire regulations, will also be happy to advise on the necessary precautions.

Radiation

Electro-magnetic radiation ranges from radio waves at one end of the spectrum to ultraviolet light and gamma rays at the other.

Employers should endeavour to reduce exposure of their employees to all kinds of radiation. The complex issues involved in radiation cannot be dealt with in depth in this book, but it is important for employers to identify possible sources of radiation. If your employees will be exposed to radiation, you must refer to the Ionising Radiations Regulations 1985 which cover all work with ionising radiations, including exposure to naturally occurring radon gas. The Ionising Radiations (Outside Workers) Regulations 1993 require special arrangements to protect outside workers.

All hazardous areas should also be marked and equipment should be maintained to minimise exposure. Employers should be advised of the need to wear protective equipment, for example, they should take care to avoid ultraviolet light by wearing suitable clothing and eye protection; if welding, special goggles or a face screen should be worn.

Ladders

Many industrial accidents occur when a ladder is used for a job when a tower scaffold or mobile access platform would have been safer and more efficient. Ladders are, of course, a common and useful means of access and are often used for short jobs. However, many ladder accidents happen during work lasting less than 30 minutes.

Legal requirements

The law in this area is governed by the Provision and Use of Work Equipment Regulations 1998 (PUWER), the Construction (Health and Safety Welfare) Regulations 1996 and the Construction (Design and Management) Regulations 1994.

Risk assessment

At the outset, the employer should consider whether or not a ladder is required or if there is a better, or safer, means of access. If ladders are to be used, precautions must be taken.

- A ladder can be used safely only if the person can reach the workstation from a position of one metre below the top of the ladder, i.e. there should be no undue stretching.

- The ladder must be strong enough for whatever job is to be done. The rungs must not be cracked or missing; makeshift or homemade ladders should be avoided.

- A rope or other suitable stabilisation device should secure ladders. If neither is available, it should be 'footed' by a colleague.

- Ladders should be angled so that the bottom will not slip outwards – four units up to each one out from the base.

- The top of the ladder should also be rested against a solid surface. The base of the ladder should be placed on a level and reliable surface. In a recent case, a claimant succeeded in recovering damages against his employer when he slipped whilst descending a stepladder. The surface of the floor was slippery and this had led to the soles of the claimant's shoes becoming wet. This is why he slipped from the ladder.

- Any tools should be carried in a shoulder bag or holster attached to a belt so that both hands are free.

- Do not carry heavy items or long lengths of material up a ladder.

- Ladders should not be painted as this may hide any defects.

- A good handhold should be available.

- Stepladders can be easily overturned. Do not use the top of a stepladder to work from unless it has specially designed handholds. Never overreach.

Machinery

The dangers of machinery and the number of accidents caused by dangerous machinery cannot be over-emphasised. Many cases involve young or inexperienced workers, where accidents could have been avoided through proper training.

The technical aspects of assessing and removing the risks of dangerous machinery is a vast subject and can only be dealt with briefly in this book.

There are multiple standards, both national and international, which should be used for guidance.

Legal requirements

The HSE prosecutes cases involving serious personal injury caused by machinery on a daily basis. These can result in hefty fines. In 2002, a cleaning company was fined £200,000 when an employee's arm passed through a running nip causing severe injuries.

The law in this area is largely contained in the Provision and Use of Work Equipment Regulations 1998 (PUWER).

Work equipment means any machinery, appliance, apparatus or tool, in addition to any assembly of components. The courts have tended to interpret this definition very widely. It could therefore include an electric drill, ladder, tractor, power press, paving slab or even a ship. PUWER states that every employer should do the following:

- Ensure that work equipment is constructed or adapted to be suitable for the purpose for which it is provided.

- Ensure that work equipment is maintained in an efficient state, in good working order and in good repair. This means that the employer will be legally liable if the equipment is found to be defective, even if the defect was entirely latent.

- Pay attention to specific risks created by the work equipment.

- Provide health and safety information, training and written instructions.

- Ensure that specific measures are taken to prevent risks arising from dangerous parts of the machinery (e.g. guards).

- Ensure that employees are not exposed to specific hazards, such as work equipment overheating.

Lifting operations

Under the Lifting Operations and Lifting Equipment Regulations (LOLER) 1998, lifting equipment or machinery provided for work should be:

- strong and safe enough for its use;

- marked with its safe working load;

- installed and positioned to minimise risks;

- used safely;

- thoroughly examined and inspected by a competent person on an ongoing basis.

All equipment used at work for lifting and lowering loads, including attachments and accessories (except escalators), is covered by the regulations.

Risk assessment

Look for hazards

In any assessment of machinery, two factors must be considered:

- The mechanical factors, for example, design, reliability and safety devices.

- The human factors with regard to the physical operation of the machinery.

Mechanical factors

The types of hazards presented by machinery are described in detail in the British Standards and in other publications dealing with machinery safety. If the hazard could present a reasonably foreseeable risk to a person, the part of the machinery generating the hazard becomes a 'dangerous part'.

Any assessment must cover, not only the machinery and its customary operational mode, but also the maintenance, regular cleaning and care of the machine. Think about the following:

- **Traps in machinery:** if unguarded, machinery can cause a wide range of major injury accidents including hand, arm and finger amputations.

- **Entanglement hazards:** the risk of entanglement of hair, clothing or limbs is present whenever unguarded rotating parts, drills or chucks operate.

- **Contact hazards:** for example, hot surfaces, moving conveyor belts or rough surfaces.

- **Ejection hazards:** for example, metal particles, or sparks into the face or eyes.

The human factors

Think about the physical operation of the machinery, the systems of work, the potential for operator error, the routine maintenance procedures and the removal of waste. In addition:

- do not position the machine where customers or visitors may go;

- consider whether new starters or those who have changed their jobs may use the machine;

- note that workers can act foolishly, carelessly or make mistakes.

Reduce the risk of injury

There are various specific requirements for guards and other protection devices – too many to mention here. BSI's code of practice entitled *Safety of Machinery* (BS5304) is the authoritative guidance on machinery safety in the UK and is revised at regular intervals.

Briefly, it is fundamental that guards must be in place enclosing every dangerous part of the machine, as far as is practical. If it is not practical, another method should be explored, including interlocking the guards of the machine so that the machine cannot start before the guard is closed and it cannot be opened whilst the machine is working.

Also consider trip devices, overrun devices, mechanical restraints and two-hand control devices that require both hands to operate the machinery.

Extra training may be required for machine operators. Machines should all be checked after any modifications are made. The machine controls should be clearly marked to indicate what they can do and clearly marked emergency stop controls should be provided where necessary.

There are certain practical considerations that should be observed every time a person operates a machine or a potentially dangerous one. These may be listed on a checklist next to the machinery, for example:

- You must wear the appropriate protective clothing and equipment (e.g. safety glasses).

- You must know how to stop the machine before you start it.

- You must not use a machine unless you are authorised and trained to do so.

- All guards should be in position and all protective devices should be working.

- You must not endeavour to clean a machine that is in motion. Before cleaning, machines should be switched off and detached from any mains device.

- You must report any machine to the employer or supervisor if it is not functioning properly, or if there is a suggestion that it may be faulty in any way.

Maintenance

Compliance with the law requires the operation of planned maintenance programmes. These should incorporate records which identify each item of work equipment by serial number, the procedure to be undertaken, the frequency of maintenance and the individual responsible for undertaking the procedure.

Construction

HSE inspectors visited 1,446 construction sites in June 2003 and stopped work at almost a quarter during the national blitz on preventing falls from heights.

The construction industry covers a wide field of operations from a large civil engineering project such as an office block to a self-employed tradesman carrying out minor repairs to a building.

Anyone working full-time in construction must consult a more detailed source of information. However, most offices and businesses require internal and external work at some stage. The extensive employment of casual labour and the real problems that can arise when contractors are working on an existing site contribute to the high incidence of accidents in the industry.

Scaffolding and roofing work is inherently dangerous. However, other potential risks include falling cables, inadequate work platforms and faulty and unstable equipment. Ill-health can also result from exposure to dust including asbestos, which can cause respiratory diseases and cancer. In addition, hernias and foot, hand and back injuries are all commonplace.

As a result, insurance premiums in this sector are soaring and those in the construction industry often find it difficult to get insurance. In breach of legal requirements, some firms trade without employer's liability insurance which can attract fines of over £2,500 per day. A good safety record will assist a small to medium-sized enterprise in obtaining a competitive premium.

However, the risks associated with building sites do not just apply to builders. All business owners are legally required to ensure the safety of contractors whilst working on their premises.

Legal requirements

Under criminal law (the HSW Act), there is concurrent responsibility for the safety of operations. Therefore the company, which is the occupier of the premises or the site concerned, can be subject to prosecution if there are breaches of the law on its premises, even if they are caused by a contractor or subcontractor. The business owner must take every reasonably practicable measure to ensure compliance. In addition to the above, employers have duties under the following regulations:

- The Construction (Health, Safety and Welfare) Regulations 1996
- The Construction (Design and Management) Regulations 1994

- The Construction (Head Protection) Regulations 1989

The Construction (Health, Safety and Welfare) Regulations 1996 impose requirements on those carrying out the work and on those who may be affected. The principal provisions include the following:

- Precautions against falls including a specific criteria for ladders.

- Precautions against falling objects including covered walkways and the proper storage of materials and equipment.

- Safe work on structures.

- Prevention or avoidance of drowning.

- Control of traffic routes, vehicles, doors and gates.

- Prevention and control of emergencies.

- Welfare facilities, for example, sanitary, washing facilities and an adequate supply of drinking water.

- Training, inspection and written reports made by a competent person.

The Construction (Design and Management) Regulations 1994 specify the relationships that must exist between a client, principal contractor and other contractors from a health and safety perspective.

For example, a client (i.e. the person for whom a project is carried out) must appoint a planning supervisor and principal contractor in respect of each project and ensure that information is available in a health and safety file for inspection by specified people. With regard to construction projects, a health and safety plan must be prepared at the pre-tender and construction phase of the project.

The Construction (Head Protection) Regulations 1989 impose duties on employers, those in control of construction sites, the self-employed and employees regarding the provision and use of head protection whenever there is a foreseeable risk of head injury. Employers must provide and maintain head protection and ensure that it is worn. Employees also have a duty to wear head protection in designated areas and operations.

Risk assessment

Look for hazards and reduce the risk of injury

Hazards arise during even the most straightforward activities. For example, the risks associated with painting may include a risk of falling from a ladder and from inhalation of paints and solvents. Fitting out an office, on the other hand, may involve a risk of falling, electrical risks from portable equipment and exposure to asbestos.

Risk of falls

Check the following every time:

- Be especially aware of trespassers (to whom legal duties are also owed), particularly children.

- If the work is undertaken via a scaffold, platform or a ladder, it is vital to ensure that safe access onto, and off, the work area is facilitated with the proper equipment. This includes properly tied scaffolds, platforms which have guard rails and toe boards. Further guards such as brick guards may also be required to provide extra protection to prevent tools and materials from falling.

- Platforms, and any scaffolding, should be wide enough for the work to be done and they should be fully boarded. The boards themselves should be properly supported and should not overhang excessively.

- The scaffolding should be inspected regularly, at least once a week, and more regularly in inclement weather.

- For scaffolding, before starting work on any scaffold or other working platforms, an employer must follow some relatively obvious (but often ignored) safety precautions, to include the following:

  - Any equipment is installed by competent and specialist people.

  - A handover certificate is supplied by the provider of the equipment (to cover matters such as the operation and maintenance of equipment).

  - Any necessary barriers or covered walkways are provided.

- Equipment is protected from the weather if necessary.

- Appropriate notices are attached to warn not only employers or sub-contractors but also vandals or trespassers of the dangers of the equipment.

- As far as roofs are concerned, there must be a safe means of moving across the roof and onto sloping and fragile roofs. Purpose-made roof ladders or crawling boards should be supplied.

- Home-made ladders should not be used under any circumstances and have been the cause of many industrial accidents.

- Broadly similar rules apply for tower scaffolds and mobile and suspended access equipment.

- As ever, the person responsible for the scaffold or the mobile access equipment should be fully trained and the work platforms should be provided with guard rails, toe boards and other suitable barriers. They should be on firm and level ground and there should be safe access to and from the platform.

- For towers and working platforms, follow the manufacturers' instructions. Do not exceed the maximum height allowed for a given base dimension.

- In cases where it is impracticable to provide a working platform, boatswain's chairs and seats can be used. Employers should also give consideration to using safety harnesses. Whilst a harness may not prevent a fall, it will minimise the risk of serious injury.

- Even in offices it must be possible for windows to be opened and cleaned safely. You may need to fit anchor points if window cleaners have to use harnesses.

Risks from machinery and powered hand tools

- You do not work from height unless it is essential.

- Only competently trained people should deal with any specialised equipment and training should be given in the use of such equipment, as required.

- The manufacturer's instructions should be observed at all times.

Risks from electrocution

- Other common, perhaps less obvious, problems on building sites are injuries from electricity cables, gas pipes, etc. In circumstances where digging work has to be undertaken, this should take place well away from any such cables and pipes, or cable detection devices should be used.

- Employers will be expected to consider service plans, safe digging practices and manufacturers' or suppliers' briefing documentation.

- Check that there are no overhead power lines in the way or any obstructions.

Risks from collapses of excavations

- Trench sites can collapse suddenly whatever the nature of the soil. Any excavations deeper than 1.2 metres must have the sides supported.

- Dig well away from the underground services.

Vehicles in the workplace

Many deaths and serious injuries that occur in the workplace involve the use of cars, vans, heavy goods vehicles, etc. The most common types of incident involve people being struck by a moving vehicle, people falling from a vehicle, objects falling from vehicles or from vehicles overturning. However, there is also a risk of accidents due to over-tiredness. Later in this section we deal with the legal requirements for tachographs.

Legal requirements

Vehicles can be included in the definition of 'work equipment' in the Provision and Use of Work Equipment Regulations 1998 (PUWER). The legal requirements for tachographs are contained in both UK domestic law and European legislation which also applies to England and Wales. The relevant European regulation is Regulation 3821/85ECC.

Risk assessment

Look for hazards

- Any vehicles used must be assessed regularly and new vehicles should be provided when appropriate.

- Look also at those who use the vehicles and the condition in which they use them: is there a risk of overheating or the employee working long hours and becoming sleepy? The fact that the greatest risk of being killed at work is due to occupational road use has long been recognised. Employers should therefore create systems for monitoring insurance claims, accident records and damage reports. If there is a problem, then the employer should take action.

- Consider the risks to employees during the repair and maintenance of vehicles, for example, back injuries are particularly commonplace. Also consider explosion risks when draining and repairing fuel tanks.

Reduce the risk of injury

Preventative steps or precautions can be taken to minimise the risk of accidents occurring.

The vehicles

- Are there any potentially dangerous parts on any vehicles, for example, exposed exhaust pipes?

- Have seat belts been fitted?

- Have the vehicles been serviced comprehensively and regularly?

- Is there a safe means of access to and from cabs, vans or heavy goods vehicles?

- Are the drivers carrying out basic safety checks before driving the vehicles?

- Are the vehicles provided with appropriate reflectors, horns, lights, etc?

Vehicle activities

- Are there pre-employment checks on drivers' licences and their fitness to drive?

- Drivers should be given full and comprehensive guidance as to how to reverse their vehicle and park it appropriately. Employers should ensure that there are external and side mounted mirrors to facilitate the best possible all-round visibility. Also consider whether a banksman should be provided to assist with any reversing procedure. This is particularly important for larger vehicles. Claims regularly involve crushing injuries to feet as a result of reversing vehicles.

- Employers should consider whether there is a designated parking area and advise that vehicles and any trailers are securely braked before being parked and subsequently locked.

- Proper guidance should also be given as to the loading and unloading of vehicles including the following:

  - Check that the loading or unloading that is being carried out is done in an area that is even and free from obstruction.

  - Loading and unloading should take place in a safe place away from traffic and pedestrians if possible.

  - The load should be spread evenly to avoid the vehicle becoming unstable.

  - Guidance should be given as to the capacity of any vehicle so that it is not overloaded.

  - The load should be secured as far as possible.

- Consider using protective equipment to prevent, for example, the inhalation of asbestos dust from brake and clutch lining pads or to prevent contact with battery acid.

The workplace

- Employers should ensure that the vehicle routes are appropriate for the size and number of vehicles that will use the route, i.e. that they are well maintained, free from hazard, even and of the proper width.

- Consider providing speed humps or barriers to reduce vehicle speeds and to keep vehicles and pedestrians apart.

- Traffic signs should be provided if appropriate, such as 'Give Way', 'No Entry' and any speed limit signals should also be erected.

- Adequate parking should be provided and pedestrian crossings should possibly be considered. Is a one-way system appropriate, for example?

- Systems should be in place to deal with breakdowns and other emergencies, such as the use of mobile phones and safe systems of work for working on or by roads.

Tachographs

All vehicles exceeding 3.5 tonnes used for carrying goods and passengers must be fitted with an EC approved tachograph to record the number of hours driven by the driver, in addition to his other work and rest periods.

Certain vehicles are exempt from the European rules but then the UK domestic rules apply. Further information can be obtained from your local Department of Transport testing station. Generally, however, the following should be noted:

- The daily driving limit is nine hours (which can be increased to ten hours twice a week) taken between two consecutive daily rest periods. After four and a half hours of cumulative or continuous driving, a driver must take a break of at least 45 minutes.

- Tachographs must be inspected at a DOT approved tachograph calibration centre at the following times:

    - Every two years to check that the system is working properly.

    - Every six years to calibrate the tachograph.

- Drivers must carry their record charts (not photocopies) for the current week and for the last day of the previous week. Charts must be given to employers within 21 days.

- Employers must do the following:

  - Keep all charts for at least one year.

  - Make regular checks to make sure that the rules are being obeyed.

  - Pass over record charts to enforcement authorities when so requested. Failure to do so is a criminal offence.

Fourth EU Motor Directive 2000/26/EC

This European Directive came into force on 20 January 2003. The aim of the Directive is to assist those people injured in a road traffic accident in other European countries to make a claim for compensation. This system will ensure that claimants will be able to pursue the claim quickly by being able to identify the insurer of the liable vehicle from its registration number.

Each country therefore has to set up an information centre to provide insurer information from the registration number. In the UK, this is the Motor Insurers' Information Centre (MIIC) in Milton Keynes – see the Appendices for their contact details. Requests must be made in writing accompanied by a £10 fee. The MIIC will then provide the owner's address, policy number and the identity of a UK representative who will deal with the claim.

Every vehicle in circulation comes under the Directive. This does create difficulties in respect of vehicles insured on a fleet policy (where the insurer may not hold details of each vehicle) and also on motor trade policies (where there are rapid changes of ownership). Possible solutions may include a requirement on the policyholder to submit details regularly to their insurer or directly to the database.

Farming

A national farm inspection blitz carried out by the HSE in 1997 identified basic shortcomings in many farms and was a major cause for concern.

During the campaign, 122 inspectors made over 4,000 visits to farms. Inspectors used their formal enforcement powers at up to half the premises they visited across the country. In addition to approximately 1,000 enforcement notices, 14 prosecutions were considered as a result of the circumstances found during this inspection. The most common reason for immediate enforcement was inadequately guarded machinery. However, poor electrical systems, lack of adequate chemical storage, lack of training in a variety of areas and lack of overall risk assessments were also found.

In short, agriculture continues to be the most hazardous industry to work in. Anyone involved in farming should seek separate detailed advice. Many of the previous regulations referred to apply and you should also refer to the previous relevant sections, including section 1 Slips/trips, section 2 Manual handling, section 5 Hazardous substances, section 6 Workers away from base, section 8 Stress, section 9 Noise, section 10 Vibration, section 11 Electricity, section 13 Machinery and section 15 Vehicles in the workplace. In addition, specific regulations relating to agricultural businesses will apply.

CHAPTER 4

Accidents at work

Even the company which follows the health and safety law to the letter will experience an accident in the workplace at some time or other. It may only be a secretary with a paper cut requiring a plaster or a bruise following a trip down a staircase, but when injuries and accidents occur, you must be able to deal with them. If an incident does occur, then you should follow the checklist below.

Checklist

Take the action required to deal with the immediate risk

If first aid is required, ensure that the first aid box is replenished.

Complete the accident book

This should be completed for all accidents at work.

RIDDOR form

The Reporting of Injuries, Diseases and Dangerous Occurrences Regulations 1995 (RIDDOR) requires employers, the self-employed or those in control of workplaces to report some work-related accidents, diseases and dangerous occurrences to the Health and Safety Executive (HSE). Reporting of accidents can be made via the internet, phone, fax and email or by post to the Incident Contact Centre in Caerphilly – see the Appendices for their contact details.

Which accidents must I report?

- **Death or major injury:** if an employee, self-employed person or person working on your premises is killed or suffers a major injury, or a member of the public is killed in an accident connected with work, you must notify the HSE or your local authority immediately. Within ten days, you must follow this up by completing an accident form (Form F2508) – see sample copy on pages 110–11.

- **Over three day injury:** if an accident connected with work results in an employee or a self-employed person working on your premises having three days away from work or being unable to do the full range of duties, then Form F2058 must be completed and sent to the HSE or your local authority (depending on the appropriate enforcing authority – see chapter 1) within ten days.

- **Disease:** if a doctor notifies you that an employee suffers from a reportable work-related disease, then you must complete Form F2508 and send it to the HSE or your local authority within ten days. Diseases include:
  - certain poisonings
  - occupational dermatitis
  - skin cancer
  - chrome ulcer
  - occupational asthma
  - pneumoconiosis
  - asbestosis
  - mesothelioma

- hepatitis
- anthrax
- tetanus
- hand/arm vibration syndrome

This list is not exhaustive and a full list of reportable diseases can be obtained from the HSE's Infoline. (See the Appendices for useful telephone numbers and a list of the most common occupational diseases and their causes.)

Accident investigation

It is essential to ascertain the cause of an accident and, once the cause has been established, to prevent a recurrence of that accident. Consider whether you need to take witness statements and record other information for future use, for example, in case there may be a prosecution or a civil claim. See sections 4 and 5 for further information.

Find out what happened and why. Also look at near misses, as often it is only by chance that someone was not injured. Most accidents have more than one cause so try and deal with the root causes.

Post-accident risk assessment

This should be considered in any case and certainly undertaken in the case of a serious or recurring accident.

Tell your insurer

It will be a term of your policy that certain accidents must be reported. Check your own employer's liability insurance policy for details. See also section 4 on defending a civil claim.

RIDDOR Form F2508

Health and Safety at Work etc Act 1974
The Reporting of Injuries, Diseases and Dangerous Occurrences Regulations 1995

HSE
Health & Safety Executive

Report of an injury or dangerous occurrence

Filling in this form
This form must be filled in by an employer or other responsible person.

Part A

About you

1 What is your full name?

2 What is your job title?

3 What is your telephone number?

About your organisation

4 What is the name of your organisation?

5 What is its address and postcode?

6 What type of work does the organisation do?

Part B

About the incident

1 On what date did the incident happen?

2 At what time did the incident happen?
(Please use the 24-hour clock eg 0600)

3 Did the incident happen at the above address?

Yes ☐ Go to question 4

No ☐ Where did the incident happen?

☐ elsewhere in your organisation – give the name, address and postcode
☐ at someone else's premises – give the name, address and postcode
☐ in a public place – give details of where it happened

If you do not know the postcode, what is the name of the local authority?

4 In which department, or where on the premises, did the incident happen?

F2508 (05.00)

Part C

About the injured person

If you are reporting a dangerous occurrence, go to Part F. If more than one person was injured in the same incident, please attach the details asked for in Part C and Part D for each injured person.

1 What is their full name?

2 What is their home address and postcode?

3 What is their home phone number?

4 How old are they?

5 Are they

☐ male?
☐ female?

6 What is their job title?

7 Was the injured person (tick only one box)

☐ one of your employees?
☐ on a training scheme? Give details:

☐ on work experience?
☐ employed by someone else? Give details of the employer:

☐ self-employed and at work?
☐ a member of the public?

Part D

About the injury

1 What was the injury? (eg fracture, laceration)

2 What part of the body was injured?

RIDDOR Form F2508 (continued)

3 Was the injury (tick the one box that applies)
- [ ] a fatality?
- [ ] a major injury or condition? (see accompanying notes)
- [ ] an injury to an employee or self-employed person which prevented them doing their normal work for more than 3 days?
- [ ] an injury to a member of the public which meant they had to be taken from the scene of the accident to a hospital for treatment?

4 Did the injured person (tick all the boxes that apply)
- [ ] become unconscious?
- [ ] need resuscitation?
- [ ] remain in hospital for more than 24 hours?
- [ ] none of the above.

Part E

About the kind of accident
Please tick the one box that best describes what happened, then go to Part G.

- [ ] Contact with moving machinery or material being machined
- [ ] Hit by a moving, flying or falling object
- [ ] Hit by a moving vehicle
- [ ] Hit something fixed or stationary

- [ ] Injured while handling, lifting or carrying
- [ ] Slipped, tripped or fell on the same level
- [ ] Fell from a height
  How high was the fall? _____ metres

- [ ] Trapped by something collapsing

- [ ] Drowned or asphyxiated
- [ ] Exposed to, or in contact with, a harmful substance
- [ ] Exposed to fire
- [ ] Exposed to an explosion

- [ ] Contact with electricity or an electrical discharge
- [ ] Injured by an animal
- [ ] Physically assaulted by a person

- [ ] Another kind of accident (describe it in Part G)

Part F

Dangerous occurrences
Enter the number of the dangerous occurrence you are reporting. (The numbers are given in the Regulations and in the notes which accompany this form)

Part G

Describing what happened
Give as much detail as you can. For instance
- the name of any substance involved
- the name and type of any machine involved
- the events that led to the incident
- the part played by any people.

If it was a personal injury, give details of what the person was doing. Describe any action that has since been taken to prevent a similar incident. Use a separate piece of paper if you need to.

Part H

Your signature
Signature

Date

Where to send the form
Incident Contact Centre, Caerphilly Business Centre, Caerphilly Business Park, Caerphilly, CF83 3GG. or email to riddor@natbrit.com or fax to 0845 300 99 24

For official use
Client number   Location number   Event number
[ ] INV REP [ ] Y [ ] N

HSE investigations

As we have discussed, the health and safety law is enforced by inspectors from the HSE or by local authority inspectors. They may visit at any time, not only after an accident at work, and they have the right to enter any workplace without having to give notice.

What will happen if the inspector finds a breach?

The action taken by the inspector will depend upon the nature of the breach. He may deal with the matter in the following ways:

Informally

If the breach is minor, he may simply tell the employer how to comply with the law and explain why it is necessary.

An improvement notice

If the breach is more serious, an improvement notice may be issued. This tells the employer to do something positive in order to comply with the law. After serving the notice, the inspector will discuss with the employer what needs to be done, why and when. This will also be outlined in the notice.

The notice also outlines the period within which to undertake the action. It must be at least 21 days. An inspector can take further legal action if the notice is not complied with within the specified time period but the employer may also appeal to an industrial tribunal once the notice is served if they are not happy with it.

A prohibition notice

If an activity involves a risk of serious personal injury, a prohibition notice may be served prohibiting the activity immediately or after a specified period of time. Such a notice will direct that specified activities should not be carried out unless certain remedial measures have been complied with. It is served where there is an immediate threat to life and in anticipation of danger.

An employer may also appeal a prohibition notice. However, unlike an appeal against an improvement notice where the notice is automatically suspended, the requirements of the notice continue to apply. Appeals are dealt with by industrial tribunals. Contact should be made with your local HSE who will provide you with further information on appealing.

Prosecution

In some cases, a prosecution may be initiated. This is frequently the outcome of an employer failing to comply with an improvement or prohibition notice but an inspector can also institute proceedings without serving a notice. Cases are normally heard in a Magistrates' Court but there is provision for a case to be heard in the Crown Court. Much depends upon the gravity of the offence. The court has considerable scope for punishing offenders and deterring others. In more serious cases, there are unlimited fines and, in some cases, prosecution can result in imprisonment.

However, the HSE are unlikely to prosecute for minor breaches of the law and will often give the employer the opportunity to explain the circumstances and give their views. The employer's attitude to health and safety and the safety record of the business in general are taken into account when they are deciding whether or not to prosecute.

Particular factors, taken into account when deciding whether to prosecute, are set out in the HSE's Enforcement Policy Statement. This can be obtained from the HSE. For more detailed information regarding HSE prosecutions, please refer to section 5.

Corporate liability

A company is liable when it commits an offence which is proven to have been committed with the consent or connivance of the company or is attributable to the neglect on the part of a director, manager, secretary or other similar officer. Both the individual(s) involved as well as the company may be prosecuted and found guilty of an offence. See section 6 for further information.

This means that a company director can be prosecuted even though the act or omission resulting in an offence was committed by a junior official or executive.

How a civil claim works

Despite concerns about a 'claims culture', a surprisingly small percentage of accidents leads to claims. However, it is helpful to understand the basic rules applicable to personal injury claims, whether you are a potential defendant or claimant.

Liability

For a personal injury claim to succeed, it must be proved that someone else was negligent or in breach of statutory duty, and that the negligence or breach caused the injuries and other losses. Negligence is careless conduct causing damage to others. As stated previously, a breach of regulations does not necessarily confer a right of action in civil proceedings. Specialist legal advice should always be sought from a personal injury solicitor approved by the Law Society.

The claimant must show that the defendant owed a duty of care to the claimant. An employer clearly owes a duty of care to his employee. The claimant also has to establish that the defendant has not taken care of him and that the claimant has suffered damage as a result. A negligent act can be a positive action, or simply an omission to do something. If a claimant was also partly to blame for an accident ('contributorily negligent'), his damages will be reduced by the appropriate percentage.

Who can claim?

Someone who has suffered a physical injury can obviously claim. More technical rules govern claims for psychiatric injury, such as post-traumatic stress disorder (PTSD). The courts are concerned with not 'opening the floodgates' of PTSD claims and strict rules have accordingly been developed which limit, sometimes rather artificially, who may bring a claim for damages for PTSD.

If someone has died as a result of an accident or disease, the deceased's claim for damages for personal injuries and other losses may be pursued through his estate. Any dependants may have a claim for loss of support. A bereavement award, now £10,000, may also be claimed by the deceased's spouse or a minor's parents (subject to some further limitations).

What can be claimed?

A claim is usually separated into general damages, special damages and future loss. Interest is payable as well.

- **General damages:** include damages for pain and suffering, i.e. compensation for the injury itself. This is often a surprisingly small proportion of a large award. The largest awards are usually made where there are substantial claims for loss of earnings or the cost of care – see below.

- **Special damages:** items of past loss which are capable of being calculated reasonably accurately. They include loss of earnings, travelling expenses for medical treatment, cost of medical treatment and medicines and personal effects damaged in an accident.

- **Future loss:** the calculations can be quite complex. Essentially, if there are losses that are likely to continue into the future, the claimant will not just recover the annual loss multiplied by the number of years for which the loss is likely to last. A multiplier is applied, which takes into account a discount to adjust for the fact that the claimant will be receiving the money all at once rather than over a period of years (and can therefore invest it and obtain interest on it) as well as for the other uncertainties of life.

- **Interest:** payable at different rates depending on the type of damages. Interest on special damages runs from the date of the accident, and interest on general damages runs only from the date of the service of proceedings.

- **Benefits:** received from the state by the claimant and may be deducted from the damages at the end of the case. The Compensation Recovery Unit (CRU) calculates the benefits received and supplies a certificate to the defendant (or more usually his insurers). The defendant then has to pay the CRU the amount demanded when the case is settled and will then pay the claimant's damages net of that deduction. Not all the benefits are necessarily deducted from the damages; complicated rules apply.

- **Provisional damages:** once a claim is settled, the claim has ended and no further claims can be brought. An exception to this rule is if provisional damages are awarded. This may occur if there is, at the time of settlement, a chance of significant future deterioration in the claimant's condition. A common example is when, following a head injury, there is a small increased risk of developing epilepsy. The claimant will be given the opportunity to make a further claim within a specified period should the deterioration occur.

How a claim can be made

It is important to consult a specialist solicitor who is a member of the Law Society's Personal Injury Panel. Most solicitors will offer an initial free interview when the question of costs, as well as the likely merits of a claim, can be discussed. Citizens Advice Bureaux may provide initial advice and can also help with choosing a solicitor. The Law Society maintains a list of specialist solicitors (see the Appendices for their details).

Generally, the person who was negligent or their employers will be sued. In practice, the defendant's insurers deal with the defence of most personal injury claims. If injuries result from a criminal act, then a claim may be lodged with the Criminal Injuries Compensation Authority (see the Appendices for their contact details).

Most personal injury claims settle without the need for a court hearing. There are, however, built-in unavoidable delays in any personal injury claim. At least one medical report has to be obtained to prove the injuries. It may be necessary to wait some years before an accurate prognosis is available.

A typical case will pass through the following stages:

- **Pre-action protocol:** the parties exchange information before the proceedings are started. Normally the claimant's solicitor will send the defendant's insurers a detailed letter setting out the arguments on liability and giving a general idea of the value of the case. The defendant then has three months to respond. If the defendant then admits liability, there will be a further period of negotiation in an attempt to settle the claim.

- **Starting proceedings:** a claimant's statement of case will be served and the claimant must sign a Statement of Truth stating that the contents of the statement of case are true. If the statement of case is subsequently found to be untrue, then the claimant may be punished for contempt of court. The defendant must acknowledge the service of proceedings within 14 days, failing which the claimant may enter judgment.

- **Defence:** usually this must be served within 28 days at the latest from the service of proceedings. Extensions of time may be granted, but only if there is a good reason.

- **Allocation:** the court will decide which 'track' the case should be allocated to after the parties have filed an Allocation Questionnaire. Generally, personal injury claims, where the damages for pain and suffering are probably worth less than £1,000, will be allocated to the Small Claims Track and will be dealt with more informally. (Very limited costs will also be awarded to the successful party, so it is unlikely to be economic to instruct a solicitor.) Claims worth over £1,000 but less than £15,000 will be allocated to the Fast Track (where trials are usually limited to one day), and claims over that amount to the Multi Track.

- **Disclosure:** exchange of lists between the parties of the documents relevant to the case. The lists have to be verified by a Statement of Truth. Parties have a duty to disclose all of the documents that may be relevant to a case, even if harmful to that party's own case (see page 122 for the types of documents which must be disclosed).

- **Exchange of witness statements and expert reports:** the court will set a timetable for this. The aim is that evidence should be disclosed in good time before any trial in order to avoid time wasting in court and to maximise the chance of a negotiated settlement.

- **Case management hearing:** this is more likely to happen in a Multi Track case. The aim of the hearing is for the judge and the parties, or their legal representatives, to review the case, to explore whether some matters can be agreed and to see whether it is possible to negotiate a settlement. Mediation may be considered and is actively encouraged by the courts.

- **Trial:** this is meant to be a last resort and the parties will have been encouraged to settle before this stage is reached. A Fast Track trial will normally last only one day and expert evidence is usually in writing. Multi Track trials may last many days depending on the complexity of the case. In both cases, a trial window will have been allocated earlier on in the case so the parties should have an idea of roughly when and where the case is likely to be heard. Depending on the case, a more accurate date is given nearer the time of trial. Being a witness can be very stressful – see the checklist in the next chapter for some helpful hints.

- **Payments:** damages are usually paid by the defendant's insurer, subject to any excess applicable. In long running claims, interim payments may be obtained. These are payments on account of the eventual likely damages that may be awarded. Defendants may make a payment into court, or a 'Part 36 payment', to try and make the claimant settle. If the claimant does not beat the amount paid into court at trial, he is likely to be ordered to pay the costs of both sides from the date of the payment made. A claimant may make a Part 36 offer setting out what he would accept in settlement of his claim. A defendant who fails to beat such an offer will also be penalised in costs and will have to pay a penalty rate of interest as well.

- **Costs:** the defendant's insurer also usually pays these if the claimant is successful.

- **Limitation:** court proceedings in personal injury cases must generally be started within three years of the date of the accident. More time may be allowed, for example, if the claimant did not know of the injury at that time. In industrial disease cases, such as mesothelioma claims, claims are commonly brought many years after first exposure as the disease may only appear much later. The court also has the discretion to allow more time in some cases, but this should not be relied on if it can be avoided. Minors and people with a mental disability generally have longer. Different periods apply to different types of claims. Claimants should therefore seek legal advice as quickly as possible after they become aware of the possibility of a claim.

- **Inquests:** the purpose of an inquest is not to establish or apportion blame but to establish the cause of death. Coroners will generally prevent questions to witnesses if it appears that they are only trying to establish fault. Evidence is given on oath.

An inquest will be held when the coroner has reasonable cause to suspect that the deceased has died an unnatural or violent death or he has died a sudden death, the cause of which is unknown, or has died in prison. There will not be an inquest where serious criminal proceedings result from the death.

There may be a jury, for example, if the death was caused by an accident, disease or poisoning. Notice of these must be given to a government department or inspector under any Act, or if the death occurred in circumstances where the continuance or possible recurrence of the factor is prejudicial to the health or safety of the public.

Interested parties have the right to question witnesses at the inquest either in person or through their legal representative.

Possible verdicts at an inquest include unlawful killing, natural causes, industrial disease, lack of care, suicide, accidental death, misadventure or an open verdict.

Whilst inquests can be helpful to the bereaved as a milestone which has been passed and possibly as the source of some explanation of what happened, they can also be extremely traumatic. This may be particularly the case as some coroners ask the pathologist to go through the whole postmortem report in horrifying and unnecessary detail. Normally the coroner's officer should warn relatives in advance that they may wish to leave the court at this point, but this is not always the case.

Sources of further information

Citizens Advice Bureaux can give general advice. The Law Society can recommend suitably specialised solicitors. The latest version of the Civil

Procedure Rules, together with useful links to other sources of information, can be found on the Department for Consitutional Affairs' website at www.dca.gov.uk/civil/procrules_fin (see the Appendices).

Defending a civil claim

For the way a claim works, see the previous section. This section covers additional matters specifically relevant to defendants. Generally, an employer will have an insurer who will deal with civil claims. Employer's liability cover is compulsory. However, the employer must co-operate fully with that insurer, in particular by promptly telling the insurer of any possible claim and quickly forwarding any correspondence or proceedings to the insurer. The employer must also be open with the insurer and tell him anything that may be relevant to the claim. If the employer does not, the insurer may be entitled to refuse to pay under the policy. The insurer may pay the claimant but may seek to recover the money from the employer.

Informing the insurer

An insurer must be told promptly of any claim. This extends to potential claims, not just cases where there has been a formal claim. Different insurers will have different requirements and the employer should make sure that he is aware of the particular insurer's requirements. However, to be safe, an employer should notify his insurer, either directly or through his broker, as soon as there is a possibility of a claim, which probably means as soon as there has been a notifiable accident. The insurer will send a claim form that should be completed as fully as possible and returned promptly. If there is some information that will take a while to collate, send the form back with what information can be given straight away with a covering letter explaining what is to come and when it is hoped that you will be able to send it. The insurer can then say if this timescale is likely to pose a problem.

Ensure that a systematic note is kept of all claims references as failure to give these when writing to insurers will cause delay.

If court proceedings are issued, it is particularly important that the insurer is told straight away, as by that stage strict time limits apply and the insurer may be able to argue that its position has been prejudiced by any delay in forwarding proceedings. Fax and post the proceedings to the insurers. Sometimes, if a broker has been involved, proceedings may go to the broker first but, in this case, do ensure it is clear that it is the broker's obligation to quickly send the proceedings onto the insurers. Either way, ensure that receipt of the proceedings has been acknowledged in writing.

Investigating

An employer will have a record in his accident book of any accident. In all but the most minor cases, it is good practice to investigate any accident fully. It is often helpful to have short witness statements written by those who saw the accident or who otherwise have useful comments, such as the training given, whether usual practices were followed, or whether there was something unusual about the particular task that led to the accident.

The use for these statements is twofold. Firstly and most importantly, it enables a further risk assessment, based on all the information, to be carried out which details the action to be taken to avoid a recurrence of the accident. Secondly, it ensures that accounts of the accident are obtained whilst matters are still fresh in people's minds. Claims may materialise some years after the events in question, by which time it may be too late to take meaningful statements.

Bear in mind, however, that these statements are likely to be disclosable in subsequent civil proceedings if the employer takes them for this dual purpose. Different rules may apply if the insurer investigates an accident solely for the purpose of defending a possible civil claim. In this case, the statements may not be automatically disclosable. If the accident was a serious one, it may be wise to speak to the insurers before doing more than the initial investigations as they may wish to be involved.

Record keeping

A wide range of documents may be relevant to an accident claim and disclosable to a claimant, depending on the circumstances. The claimant may request to see these before the proceedings are issued and is likely to be entitled to do so. Always consult your insurer before disclosing documents to a claimant.

Documents relevant to the accident should be collated and ideally kept in one place or at least a note kept in one place of where the documents may be found. It is important to ensure that these documents are kept, ideally for as long as insurance records have to be kept, i.e. 40 years if no claim has materialised and been dealt with, although it is appreciated that space may well not permit this. Documents should, in any event, be kept for a minimum of five years to allow for the normal three-year limitation period for personal injury claims plus time for any proceedings to be pursued. Again insurers may have specific requirements.

Employers are often surprised to see the documents which they are expected to disclose in a workplace claim. It gives an indication of the onerous duties upon employers and therefore the lists are reproduced in full here. Generally, the following is required:

1. An accident book entry

2. A first aider report

3. A surgery record

4. A foreman/supervisor's accident report

5. A safety representative's accident report

6. A RIDDOR report to the HSE

7. Other communications between defendants and the HSE

8. Minutes of the health and safety committee meeting(s) where the accident/matter was considered

9. A report to the DSS

10. Any documents listed above relating to any previous accident/matter identified by the claimant and relied upon as proof of negligence

11. Earnings information

Documents produced to comply with the Management of Health and Safety at Work Regulations 1999

1. A pre-accident risk assessment

2. A post-accident re-assessment

3. An accident investigation report

4. Health surveillance records in appropriate cases

5. Information provided to employees

6. Documents relating to the employees' health and safety training

Disclosure where specific regulations apply

The Workplace (Health, Safety and Welfare) Regulations 1992

1. Repair and maintenance records

2. Housekeeping records

3. Hazard warning signs or notices

The Provision and Use of Work Equipment Regulations 1998 (PUWER)

1. Manufacturers' specifications and instructions in respect of relevant work equipment establishing its suitability

2. Maintenance log/records

3. Documents providing information and instructions to employees

4. Documents provided to the employee in respect of training for use

5. Any notice, sign or document relied upon as a defence to alleged breaches of Regulations 14 to 18 dealing with controls and control systems

6. Instruction/training documents issued to comply with the requirements of Regulation 22 insofar as it deals with maintenance operations where the machinery is not shut down

7. Copies of markings required to comply with Regulation 23

8. Copies of warnings required to comply with Regulation 24

The Personal Protective Equipment at Work Regulations 1992 (PPE)

1. Documents relating to the assessment of personal protective equipment (PPE)

2. Documents relating to the maintenance and replacement of PPE

3. A record of the maintenance procedures for PPE

4. Records of tests and examinations of PPE

5. Documents providing information, instruction and training in relation to PPE

6. Instructions for use of PPE to include the manufacturers' instructions

The Manual Handling Operations Regulations 1992

1. A manual handling risk assessment

2. A re-assessment carried out post-accident

3. Documents showing information provided to the employee giving general indications related to the load and precise indications on the weight of the load and the heaviest side of the load if the centre of gravity was not positioned centrally

4. Documents relating to training in respect of manual handling operations and training records

The Health and Safety (Display Screen Equipment) Regulations 1992 (DSE)

1. An analysis of workstations to assess and reduce risks

2. Re-assessment of the analysis of workstations to assess and reduce risks following the development of symptoms by the claimant

3. Documents detailing the provision of training including training records

4. Documents providing information to employees

The Control of Substances Hazardous to Health Regulations 2002 (COSHH)

1. A risk assessment carried out

2. A reviewed risk assessment carried out.

3. Copy labels from containers used for storage handling and the disposal of carcinogenics

4. Warning signs identifying designation of areas and installations that may be contaminated by carcinogenics

5. Documents relating to the assessment of PPE

6. Documents relating to the maintenance and replacement of PPE

7. A record of maintenance procedures for PPE

8. Records of tests and examinations of PPE

9. Documents providing information, instruction and training in relation to PPE

10. Instructions for use of PPE to include the manufacturers' instructions

11. Air monitoring records for substances assigned a maximum exposure limit or occupational exposure standard

12. A maintenance examination and records of the tests of control measures

13. Monitoring records

14. Health surveillance records

15. Documents detailing information, instruction and training including training records for employees

16. Labels and health and safety data sheets supplied to the employers

The Construction (Design and Management) Regulations 1994

1. A Notification of Project Form (HSE F10)

2. A health and safety plan

3. A health and safety file

4. Information and training records

5. Records of advice from, and views of, persons at work

The Pressure Systems and Transportable Gas Containers Regulations 1989

1. Information and specimen markings

2. Written statements specifying the safe operating limits of a system

3. A copy of the written scheme of examination

4. Examination records

5. Instructions provided for the use of the operator

6. Records kept to comply with the requirements of Regulation 13

7. Records kept to comply with the requirements of Regulation 22

The Lifting Plant and Equipment (Records of Test and Examination etc) Regulations 1992

1. Records kept to comply with the requirements of Regulation 6

The Noise at Work Regulations 1989

1. Any risk assessment records

2. Manufacturers' literature in respect of all ear protection made available to the claimant

3. All documents provided to the employee for the provision of information to comply with Regulation 11

The Construction (Head Protection) Regulations 1989

1. A pre-accident assessment of head protection

2. A post-accident re-assessment

The Construction (Health, Safety and Welfare) Regulations 1996

1. A report prepared following inspections and examinations of excavations, etc

2. A report prepared following inspections and examinations of work in cofferdams and caissons

Giving evidence

This can be an alarming experience even for the best prepared witness. In a large case, the witness may have had an opportunity to meet the solicitor dealing with the case and discuss any concerns. He also may have seen the barrister instructed. In a smaller case, the employer, or a representative of the insurer, may have taken the witness statement and the first the witness sees of the legal representatives is at the court door. The following guidelines may help.

Guidance on being a witness

- **Tell the truth.** Remember that in cross-examination the advocate on the other side will be trying to make you admit things in his side's favour. Do not change your story to suit what you think helps your version of events. It will be obvious and you will lose credibility.

- **Remember that you are at court to give evidence.** You are not there to win the case and you are not on trial; you are there to assist the judge.

- **Arrive in good time.** Ideally go and see a court in action before you have to give evidence.

- **Keep calm.** Take your time.

- **Listen carefully to the questions being asked.** Take time to think before answering. You will not get a second chance.

- **Ask for a question to be repeated if you do not understand it.** The judge will understand that a lawyer's language can be confusing to the most intelligent people.

- **Speak up.** The acoustics can be poor in a court room.

- **Look at the judge when speaking.**

- **Keep your answers short.** If more information is needed, you will be asked for it.

- **Be polite.** The judge and lawyers will normally try to be as polite as possible to you. You will make a better impression if you are polite in return.

- **Keep your temper.** Do not take things personally. Concentrate on giving your evidence.

- **Ask for a copy of a document if you are being asked questions about it.**

- **Be serious.** Do not try to be funny as it is likely to backfire.

- **Dress smartly.** People in court are normally formally dressed. You do not want to feel or look out of place.

Control of the case

Generally, insurers will effectively be in control of the case and will normally have the right to make key decisions if necessary without the agreement of the insured, under the terms of the policy. If it is important to an employer that a certain line is taken on a case or if, for example, settlement of one claim may have wider implications and may lead to many others, it is important that the insurer is clearly informed of this at an early stage.

The employer must co-operate with his insurers, allow time for witnesses to be interviewed and do his best to ensure that witnesses attend court when required.

Defending HSE prosecutions

From 2003 to 2004, the Health and Safety Executive (HSE) investigated over 28,000 incidents and, as a result, issued 13,500 notices, bringing approximately 1,000 prosecutions. Of the 982 prosecutions, 89 per cent resulted in a conviction.

In recent years, there have been large increases in the level of financial penalties following successful prosecutions. The average fine per case between 2003 to 2004 increased to £13,947. Companies are more likely than ever to be prosecuted, to be convicted and to receive significant fines.

Accidents do happen and when they do, the difference between a company being prosecuted or simply being issued with a warning will depend upon whether proper care and attention has previously been given to the implementation of a thorough, well-considered, health and safety policy. This would include safe systems of work, safe plant, equipment and machinery, the adequate supervision of employees and properly conducted risk assessments.

However, in the event that the HSE do seek to prosecute a company (or indeed an individual), there are steps which can be taken to limit the extent of the financial penalties that typically follow convictions in such proceedings.

Legal advice

It is vital to obtain legal advice at an early stage. Your insurance company may cover you for the legal costs of defending a health and safety prosecution. Ensure that your solicitors have specialist experience in health and safety prosecutions.

Advance disclosure

Before a company has to inform the Magistrates' Court whether or not it is going to plead guilty or not guilty to a particular charge, the company is entitled to the disclosure of detailed information setting out the case against it, including the relevant documents and statements.

Always consider this information first before deciding on how you, or your company, intend to plead; your legal advisers will be able to assist you with this decision.

Which court?

Once you, or the company, has informed the Magistrates' Court whether you intend to plead guilty or not guilty, the magistrates will determine whether the case will be heard in the Magistrates' Court or in the Crown Court. It is important to obtain legal advice as to which court your case should be heard in.

Defence/mitigation

Your legal adviser will inform you how to defend yourself against an HSE prosecution or at least reduce the penalty imposed. However, do bear the following in mind:

- It is important to demonstrate to the court all, or any, of the company's efforts, since the day of the accident, to address any deficiencies in the company's health and safety policy.

- All health and safety documentation should be produced including copies of accident reports, any communications with the HSE and any risk assessments, as well as the company's health and safety policy.

- A health and safety consultant may assist the company in assessing the workplace after the accident and may make appropriate suggestions as to how systems can be improved. Evidence can then be provided to the court as to the timescale within which you, as employer, propose to implement the health and safety consultant's recommendations.

- As with any criminal proceedings, a level of remorse is usually appreciated by the court and any admissions, or partial admissions, will often reduce the level and nature of the penalty.

- Properly audited accounts for the previous three to five years should be provided to the court in the event that a fine is considered.

Fines

The maximum fine that can be imposed on both corporate bodies and/or individuals in the Magistrates' Court is £20,000 per offence. It should be remembered that there is no limit on the aggregate fine that can be imposed in respect of a number of different offences. In the Crown Court there can be an unlimited fine and/or up to two years' imprisonment for some offences (listed in Sections 33(4) of the HSW Act) and/or £200 per day for failing to comply with a notice.

Gone are the days when a company could expect to receive an inconsequential fine, now fines are greater and the approach is far more consistent between the courts. However, each and every case must be dealt with according to its own particular circumstances.

The courts consider a number of factors in every case:

1. In assessing the gravity of the breach, the court must consider how far short the defendant fell from doing what was reasonably practicable.

2. Death resulting from any breach (regardless of how minor the breach) must be regarded as an aggravating factor.

3. It is considered a serious aggravating factor if the defendant has deliberately profited from a failure to take necessary health and safety measures, or ran a risk to save money.

4. The degree of risk and the extent of the danger created by the offence must be considered in every case.

5. The court will consider the extent of the breach, i.e. whether it was an isolated incident or if it had been continuing for a period of time.

6. Failure to heed warnings is considered to be an aggravating factor. Such warnings range from formal advice from the HSE, to the occurrence of similar incidents in the past, or even events which might have given warnings to the employer.

7. In the company's favour the court will take into account the following:

    (a) Whether a prompt admission of responsibility was made and a quick plea of guilty.

(b) What steps were taken to remedy deficiencies after they were drawn to the defendant's attention.

(c) A good previous safety record.

The level of fine should not only reflect the gravity of the offence but also the means of the offender and this applies as much to companies as to individuals.

If the defendant company wishes to make any submissions to the court about the appropriate level of the fine and its ability to pay such a fine, it should supply copies of its accounts and any other financial information on which it intends to rely, in good time before the court hearing, to both the court and to the prosecution.

Prosecution costs

Where a person or company is convicted of an offence, the court may make a costs order against the defendant. Costs are more of an issue in health and safety cases than in other criminal cases, as the costs incurred can be much higher. However, costs orders should never exceed a sum which the company or individual is able to pay, having regards to his means and to any fine imposed on him.

Corporate manslaughter

Corporate manslaughter is a crime that can be committed by a company in relation to a work-related death. However, there has been reluctance by both the government and the courts over the last 20 years to define clearly when companies should be held responsible for their acts or omissions when a death occurs in the workplace. Whilst legislation has been in place to deal with individual acts of criminality or negligence, generally speaking, companies can only be convicted of gross negligence when a director or senior manager – a 'controlling mind and will' of the company – is found guilty of manslaughter.

Inevitably, in business today, it is the smaller companies who are more exposed to successful prosecutions. In larger companies with comprehensive health and safety systems in place, it is more unlikely that the directors will be rendered directly responsible for acts or omissions which lead to a death. Large companies will often delegate safety decisions to managers lower down the heirarchy or to outside companies. This means that generally only small organisations have been prosecuted. A recent example resulted from the Lyme Bay canoeing tragedy, R V Kite and OLL Ltd. Kite was the sole director of the company and, as a result, he was the person responsible for devising, instituting, enforcing and maintaining a suitable health and safety policy.

Briefly, in this case, two instructors, 80 students and a teacher were on a canoeing trip in Dorset. They were covering a relatively short distance (one to two miles). Not long after they commenced, some of the group drifted out to sea and the canoes became swamped. Four of the students drowned. Because Kite was the 'directing mind', he was charged with corporate manslaughter. This was the first ever successful conviction for corporate manslaughter. Kite was sentenced to three years' imprisonment, reduced to two years on appeal.

Public pressure to redress the imbalance between smaller companies and large corporations has also risen after such incidents as Zeebrugge, Piper Alpha, Southall and Kings Cross. These tragic events did not lead to the successful prosecutions of the corporate bodies involved, despite clear indications that there were safety management shortcomings.

The government therefore announced in May 2003 that it would honour its 1997 manifesto commitment to introduce a Bill on corporate manslaughter. Assuming the government adopts the proposals of the 1996 Law Commission, there will be a special offence of 'corporate killing' where 'the defendant's conduct in causing the death falls far below what should be reasonably expected'. Death will be regarded as having been caused by the conduct of a corporation 'if it is caused by a failure in the way in which the corporation's activities are managed, or organised, to ensure the health and safety of persons employed in or affected by those activities'.

Points for action

In order to avoid liability, do think about taking the following action:

- Consider appointing a director responsible for health and safety as well as a committee to assist.

- Review all working practices and ensure that the company has in writing:

  - all the relevant method statements and risk assessments;

  - all the training records for the employees;

  - all records for the installation and servicing of equipment.

- Hold regular meetings with the health and safety director and his committee. Ensure that the fellow directors are au fait with the decisions and practices and keep records of such meetings.

- Where an accident occurs, inform the company's insurers immediately.

Appendices

Useful addresses

Association of British Insurers (ABI)

Employers' Liability Enquiry Unit	Tel: 020 7600 3333
51 Gresham Street	Email: info@abi.org.uk
London EC2V 7HQ	Website: www.abi.org.uk

Association of Noise Consultants

6 Trap Road	Tel: 01763 852 958
Guilden Morden	Email: mail@association-of-
Royston	noise-consultants.co.uk
Herts SG8 0JE	Website: www.association-of-
	noise-consultants.co.uk

British Red Cross

44 Moorfields London EC2Y 9AL	Tel: 0870 170 7000 Email: information@redcross. org.uk Website: www.redcross.org.uk

British Safety Council

National Safety Centre 70 Chancellors Road London W6 9RS	Tel: 020 8741 1231 Website: www.britishsafety council.org

British Standards Institution

BSI Sales and Customer Services 389 Chiswick High Road London W4 4AL	Tel: 020 8996 9000 Email: cservices@bsi-global.com Website: www.bsi-global.com

BUPA Wellness Service

Battle Bridge House 300 Gray's Inn Road London WC1X 8DU	Tel: 0845 300 8220 Website: www.bupa.co.uk/ wellness

Centre for Stress Management

156 Westcombe Hill London SE3 7DH	Tel: 020 8293 4334 Website: www.managing stress.com

Citizens Advice

Myddelton House 115–123 Pentonville Road London N1 9LZ	Website: www.citizensadvice. org.uk

Criminal Injuries Compensation Authority

Tay House 300 Bath Street Glasgow G2 4LN	Tel: 0800 358 3601 Website: www.cica.gov.uk

Department for Constitutional Affairs

Selborne House	Tel: 020 7210 8614
54 Victoria Street	Email: general.queries@dca.
London SW1E 6QW	gsi.gov.uk
	Website: www.dca.gov.uk
	Civil Procedure Rules link: www.
	dca.gov.uk/civil/procrules_fin

Disability Rights Commission

DRC Helpline	Tel: 0845 762 2633
FREEPOST MID02164	Website: www.drc-gb.org
Stratford upon Avon CV37 9BR	

Disability Unit

Department for Work and Pensions	Benefit Enquiry Line for People
Level 6, Adelphi Building	with Disabilities: 0800 882 200
John Adams Street	Email: enquiry-disability@dwp.
London WC2N 6HT	gsi.gov.uk
	Website: www.disability.gov.uk

Financial Services Authority

25 The North Colonnade	Tel: 020 7066 1000
Canary Wharf	Helpline: 0845 606 1234
London E14 5HS	Email: consumerhelp@fsa.gov.uk
	Website: www.fsa.gov.uk

Fire Training Videos Limited

110 Warren Road	Tel: 01903 268 000
Worthing	
West Sussex BN14 9QX	

HSE Books

PO Box 1999	Tel: 01787 881 165
Sudbury	Website: www.hsebooks.com
Suffolk CO10 2WA	

HSE's First Aid at Work Approvals and Monitoring Section

Grove House   Tel: 0161 952 8322/8326/8280
Skerton Road
Trafford M16 ORB

Provides a list of over 1,600 first aid courses in the UK which are approved by the HSE.

HSE's Infoline

Caerphilly Business Park   Helpline: 0870 154 5500
Caerphilly   Email: hseinformationservices
Wales CF83 3GG   @natbrit.com
                                     Website: www.hse.gov.uk –
                                     provides free leaflets to download

Incident Contact Centre

Caerphilly Business Park   Tel: 0845 300 9923
Caerphilly   Website: www.riddor.gov.uk
Wales CF83 3GG

Institute of Occupational Safety and Health (IOSH)

The Grange   Tel: 0116 257 3100
Highfield Drive   Email: enquiries@iosh.co.uk
Wigston   Website: www.iosh.co.uk
Leicestershire LE18 1NN

Intek Training Ltd

12 Ventnor Gardens   Tel: 01582 598 005
Luton   Email: sales@intek.co.uk
Bedfordshire LU3 3SN   Website: www.intek.co.uk

International Institute of Risk and Safety Management (IIRSM)

70 Chancellors Road   Tel: 020 8600 5538/9
London W6 9RS   Email: info@iirsm.org
                                     · Website: www.iirsm.org

Law Society

The Law Society's Hall	Tel: 020 7242 1222
113 Chancery Lane	Website: www.lawsoc.org.uk
London WC2A 1PL	

Motor Insurers' Information Centre

Linford Wood House	Tel: 0870 241 6732
6–12 Capital Drive	Email: information@miic.org.uk
Linford Wood	Website: www.miic.org.uk
Milton Keynes MK14 6XT	

National Group on Homeworking

Office 26, 30–38 Dock Street	Tel: 0113 245 4273
Leeds LS10 1JF	Advice line: 0800 174 095
	Email: admin@homeworking. gn.apc.org
	Website: www.homeworking. gn.apc.org

Royal Society for the Prevention of Accidents (RoSPA)

RoSPA House	Tel: 0121 248 2000
Edgbaston Park	Email: help@rospa.com
353 Bristol Road	Website: www.rospa.co.uk
Edgbaston	
Birmingham B5 7ST	

St Andrew's Ambulance Association

St. Andrew's House	Tel: 0141 332 4031
48 Milton Street	Email: firstaid@staaa.org.uk
Glasgow G4 0HR	Website: www.firstaid.org.uk

St John Ambulance

27 St John's Lane	Tel: 0870 010 4950
London EC1M 4BU	Website: www.sja.org.uk

Stationery Office (formerly HMSO)

51 Nine Elms Lane	Tel: 0870 600 5522
London SW8 5DR	Email: customer.services@ tso.co.uk
	Website: www.tso.co.uk

Trades Union Congress (TUC)

Congress House	Tel: 020 7636 4030
Great Russell Street	Website: www.tuc.org.uk
London WC1B 3LS	

Useful publications

The following publications can be purchased on the Health and Safety Executive's website at www.hsebooks.com or on the British Standards Institution's website at www.bsonline.techindex.co.uk.

The health and safety publication abbreviations mean the following:

ACOP	Approved Code of Practice
BS	British Standard
EH	Guidance Note: Environmental health
HSE	Health and Safety Executive leaflet
HSG	Health and Safety (Guidance) booklet
INDG	Industry Advisory (General) leaflet
INDS	Industry Advisory (Special) leaflet
ISBN	International Standard Book Numbers
L	Legal series

Re: chapter 1, section 3, 'Legal requirements' (page 3)

Consulting Employees on Health and Safety: A Guide to the Law (INDG232)

Managing Health and Safety: Five Steps to Success (INDG275)

An Introduction to Health and Safety: Health and Safety in Small Businesses (INDG259REV1)

Re: chapter 2, section 2.2, 'First aiders' (page 14)

First Aid at Work: Your Questions Answered (INDG214)

Basic Advice on First Aid at Work (INDG347)

Posters: *Basic Advice on First Aid at Work* (ISBN 0717622657) and *Electric Shock: First Aid Procedures* (ISBN 0717622649) – £7.50 each

Re: chapter 2, section 2.4, 'The health and safety policy document' (page 18)

Stating Your Business: Guidance on Preparing a Health and Safety Policy Document for Small Firms (INDG324)

Fire Safety – An Employer's Guide (ISBN 0113412290)

Re: chapter 2, section 4, 'Risk assessments' (page 27)

Five Steps to Risk Assessment (INDG163)

A Guide to Risk Assessment Requirements: Common Provisions in Health and Safety Law (INDG218)

Re: chapter 2, section 8, 'Personal protective equipment' (page 36)

A Short Guide to the Personal Protective Equipment at Work Regulations 1992 (INDG174)

The Selection, Use and Maintenance of Respiratory Protective Equipment: A Practical Guide (HSG53)

Construction (Head Protection) Regulations 1989: Guidance on Regulations (L102)

Re: chapter 2, section 9, 'Signs' (page 39)

Safety signs and signals: Guidance on the Health and Safety (Safety Signs and Signals) Regulations 1996 (L64)

Re: chapter 2, section 11, 'New or expectant mothers' (page 43)

New and Expectant Mothers at Work: A Guide for Employers (ISBN 0717625834)

Re: chapter 3, section 2, 'Manual handling' (page 49)

Getting to Grips with Manual Handling (INDG143REV2)

Re: chapter 3, section 3, 'Upper limb disorders' (page 55)

Upper Limb Disorders in the Workplace (HSG60)

Aching Arms (or RSI) in Small Businesses: Is Ill-Health Due to Upper Limb Disorders a Problem in your Workplace? (INDG171)

Re: chapter 3, section 4, 'VDUs' (page 58)

Working with VDUs (INDG36REV2)

Re: chapter 3, section 5, 'Hazardous substances' (page 65)

Dust: General Principles of Protection (EH44)

A Step by Step Guide to COSHH Assessment (HSG97)

Seven Steps to Successful Substitution of Hazardous Substances (HSG110)

Health Risks Management: A Guide to Working with Solvents (HSG188)

COSHH Essentials: Easy Steps to Control Chemicals (HSG193)

Work with Asbestos Which Does Not Normally Require a Licence: Control of Asbestos at Work Regulations 2002 Approved Code of Practice and Guidance (L27)

Re: chapter 3, section 6, 'Workers away from base' (page 72)

Homeworking: Guidance for Employers and Employees on Health and Safety (INDG226)

Re: chapter 3, section 9, 'Noise' (page 81)

Sound Solutions: Techniques to Reduce Noise at Work (HSG138)

Reducing Noise at Work: Guidance on the Noise at Work Regulations 1989 (L108)

Noise at Work: Advice for Employers (INDG362)

Re: chapter 3, section 10, 'Vibration' (page 85)

Vibration Solutions: Practical Ways to Reduce the Risk of Hand-Arm Vibration Injury (HSG170)

Health Risks From Hand-Arm Vibration: Advice for Employees and the Self-Employed (INDG126REV1)

Re: chapter 3, section 11, 'Electricity' (page 86)

Electricity at Work: Safe Working Practices (HSG85)

Re: chapter 3, section 13, 'Machinery' (page 91)

Safe Use of Machinery (PD 5304:2000)

Safety of Machinery. Basic Concepts, General Principles for Design (BS EN ISO 12100-1&2:2003)

Re: chapter 3, section 14, 'Construction' (page 95)

Health and Safety in Construction (HSG150)

A Guide to Managing Health and Safety in Construction (ISBN 0717607550)

Re: chapter 3, section 15, 'Vehicles in the workplace' (page 100)

Managing Vehicle Safety at the Workplace: A Leaflet for Employers (INDG199)

Statutes

Employer's Liability (Compulsory Insurance) Act 1969

Fire Precautions Act 1971

Health and Safety at Work Act 1974

Occupiers' Liability Act 1957

Trade Union Reform and Employment Rights Act 1993

Regulations

The Construction (Design and Management) Regulations 1994

The Construction (General Provisions) Regulations 1961

The Construction (Head Protection) Regulations 1989

The Construction (Health, Safety and Welfare) Regulations 1996

The Control of Substances Hazardous to Health Regulations 2002 (COSHH)

The Electricity at Work Regulations 1989

The Employer's Liability (Compulsory Insurance) Regulations 1998

The Fire Precautions (Workplace) Regulations 1997

The Health and Safety (Consultation with Employees) Regulations 1996

The Health and Safety (Display Screen Equipment) Regulations 1992 (DSE) – Guidance on Regulations (L26)

The Health and Safety (First Aid) Regulations 1981

The Health and Safety Information for Employees Regulations 1999

The Health and Safety (Safety Signs and Signals) Regulations 1996

The Ionising Radiations (Outside Workers) Regulations 1993

The Ionising Radiations Regulations 1985

The Lifting Operations and Lifting Equipment Regulations 1998 (LOLER) – Approved Code of Practice and Guidance (L113)

The Lifting Plant and Equipment (Records of Test and Examination etc) Regulations 1992

The Management of Health and Safety at Work (Amendment) Regulations 1994

The Management of Health and Safety at Work Regulations 1999

The Managing Construction for Health and Safety (CDM Regulations) 1994 – Approved Code of Practice (L54)

The Manual Handling Operations Regulations 1992 – Guidance on Regulations (L23)

The Noise at Work Regulations 1989

The Personal Protective Equipment at Work Regulations 1992 (PPE) – Guidance on Regulations (L25)

The Pressure Systems and Transportable Gas Containers Regulations 1989

The Provision and Use of Work Equipment Regulations 1998 (PUWER)

The Reporting of Injuries, Diseases and Dangerous Occurrences Regulations 1995 (RIDDOR) – Guidance on Regulations (L73)

The Safe Use of Work Equipment Provision and Use of Work Equipment Regulations 1998 – Approved Code of Practice and Guidance on Regulations (L22)

The Safety Representatives and Safety Committees Regulations 1977

The Working Time Regulations 1998

The Workplace (Health, Safety and Welfare) Regulations 1992 – Approved Code of Practice and Guidance on Regulations (L24)

Work-related diseases and their causes

Diseases/ Conditions	Possible Causes	Description of Disease/Condition
Alveolitis	Exposure to moulds, fungal spores or heterolgous proteins during work in: 1. agriculture, horticulture, forestry, cultivation of edible fungi or malt-working; 2. loading, unloading or handling mouldy vegetable matter or edible fungi whilst it is being stored; 3. caring for or handling birds.	Inflammation of the alveoli of the lungs caused by an allergic reaction. When caused by infection, it is called pneumonia, when caused by a chemical/ physical agent, it is called pneumonitis. Farmer's Lung – Form of this caused by inhalation of dust from mouldy hay/ straw.
Angiosarcoma	Work in or about machinery or apparatus used for the polymerisation of vinyl chloride monomer.	A highly malignant tumour.

Diseases/ Conditions	Possible Causes	Description of Disease/Condition
Anthrax	1. Work involving handling infected animals, their products or packaging containing infected material; or 2. Work on infected sites.	A usually fatal, acute infection.
Asbestosis	Caused by inhalation of mainly blue or brown asbestos dust, either during mining or quarrying, or in one of the many industries in which it is used (e.g. in the making of paper, cardboard and brake linings, and as an insulating material).	Widespread scarring in the lungs leading to severe breathing difficulties.
Barotrauma	Work involving breathing gases at increased pressure (including diving).	Non-infective inflammatory changes produced in the ear.
Brucellosis	Work involving contact with: 1. animals or their carcasses; or 2. laboratory specimens or vaccines of or containing brucella.	Undulating fever, drenching sweats, pains in joints and back and headache.
Bursitis	Physically demanding work causing severe or prolonged friction or pressure at or about the knee.	Inflammation of the knee or shoulder.
Byssinosis	The spinning or manipulation of raw or waste cotton or flax or the weaving of cotton or flax.	Pneumoconiosis/ chronic inflammatory thickening of lung tissue due to inhalation of dust.

Diseases/ Conditions	Possible Causes	Description of Disease/Condition
Carcinoma	Any occupation in: 1. glass manufacture; 2. sandstone tunnelling or quarrying; 3. the pottery industry; 4. metal ore mining; 5. slate quarrying or slate production; 6. clay mining; 7. the use of siliceous materials as abrasives; 8. foundry work; 9. granite tunnelling or quarrying; or 10. stone cutting or masonry.	Any cancer.
Carpal Tunnel Syndrome	Work involving the use of hand-held vibrating tools. Is also constitutional.	Attacks of pain and tingling in the first three or four fingers of one/both hands, usually at night. Caused by pressure on the median nerve.
Cataract	Work involving exposure to electromagnetic radiation (including radiant heat).	An opacity of the lens. A cataract may be the result of a penetrating or blunt injury, infra-red energy, electric shock or ionizing radiation.
Chrome Ulcer	Caused by chromic acid which is used in several industries, particularly in chromium plating.	Deep ulcer, particularly of the nasal septum and knuckles.

Diseases/ Conditions	Possible Causes	Description of Disease/Condition
Dysbaric Osteonecrosis	Work involving breathing gases at increased pressure (including diving).	Death of the mass of bone (osteonecrosis).
Legionellosis	Work on or near cooling systems which are located in the workplace and use water; or work on hot water service systems located in the workplace which are likely to be a source of contamination.	Form of pneumonia due to bacterium distributed in nature and commonly found in surface water and soil. Aches and pains followed rapidly by a rise in temperature, shivering attacks, coughing and shortness in breath.
Leptospirosis	1. Work in places which are, or are liable to be, infested by rats, fieldmice, voles or other small mammals; 2. Work at dog kennels or involving the care or handling of dogs; or 3. Work involving contact with bovine animals or their meat products, or pigs or their meat products.	Disease caused by a group of micro-organisms normally found in rodents and other small mammals. Varies from a mild influenza-like illness to a fatal form of jaundice due to severe liver disease. Kidneys often involved.
Lyme Disease	Work involving exposure to ticks (including, in particular, work by forestry workers, rangers, dairy farmers, gamekeepers and other persons engaged in countryside management).	Comprises of arthritis associated with skin rashes, fever, encephalitis or inflammation of the heart. Transmitted by tick bites.

Diseases/ Conditions	Possible Causes	Description of Disease/Condition
Mesothelioma	1. The working or handling of asbestos or any mixture of asbestos; or 2. The manufacture or repair of asbestos textiles or other articles containing or composed of asbestos.	Malignant tumour of the pleura, the membrane lining of the chest cavity. May be asymptomatic or cause pain/coughing/ breathing trouble.
Peripheral Neuropathy	Work involving the use or handling of, or exposure to, the fumes of or vapour containing n-hexane or methyl n-butyl ketone.	Disease affecting the nerves.
Pneumoco-niosis	1. The mining, quarrying or working of silica rock or the working of dried quartzose sand, or any dry deposit or residue of silica or any dry mixture containing such materials; or 2. The breaking, crushing or grinding of flint; or 3. Sand blasting by means of compressed air with the use of quartzose sand or crushed silica rock or flint; or 4. Working in a foundry; or 5. The manufacture of china or earthenware; or 6. The grinding of mineral graphite; or 7. The dressing of granite or any igneous rock by masons; or	General name applied to a chronic form of inflammation of the lungs.

Diseases/ Conditions	Possible Causes	Description of Disease/Condition
	8. Working underground in any mine.	
Q Fever	Work involving contact with animals, their remains or their untreated products.	Fever, severe headache and often pneumonia. Disease due to organism 'Coxiella Burneti'.
Streptococcus	Work involving contact with pigs infected with streptococcus.	Variety of bacterium.
Tennis Elbow	Any job involving the jarring/straining of the elbow.	Mild inflammation in the tendons or muscles in the elbow.
Tenosynovitis	Any job involving jarring/straining/repetitive movements.	Inflammation of a tendon.
Tinnitus	Any job involving exposure to noise.	Noise heard in the ear is a frequent accompaniment of deafness caused by damage to the auditory pathway.

Index

MORE PRODUCTS AVAILABLE FROM **LAWPACK**

Personnel Manager

A book of more than 200 do-it-yourself forms, contracts and letters to help your business manage its personnel records. Areas covered include recruitment and hiring, employment contracts and agreements, handling new employees, personnel management, performance evaluation and termination of employment.

Code B417 | ISBN 1 904053 23 8 | Paperback | A4 | 268pp | £19.99 | 3rd edition

Ready-Made Company Minutes & Resolutions

This book is what every time-pressed record-keeper needs. Maintaining good, up-to-date records of meetings and resolutions is not only good practice but also a legal requirement. This book of forms makes compiling minutes of board and shareholder meetings straight-forward. It includes more than 125 commonly-required resolutions and minutes to save you time and effort.

Code B416 | ISBN 1 904053 73 4 | Paperback | A4 | 192pp | £14.99 | 3rd edition

Tax Answers at a Glance 2005/06

We all need to get to grips with the array of taxes now levied by the government. Compiled by award-winning tax experts and presented in question-and-answer format, this handbook provides a useful and digestible summary of Income Tax, Capital Gains Tax, Inheritance Tax, pensions, self-employment, partnerships, Corporation Tax, Stamp Duty/Land Tax, VAT, and more.

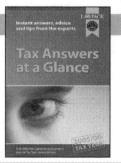

Code B425 | ISBN 1 904053 76 9 | Paperback | 153 x 234mm | 240pp | £9.99 | 5th edition

To order, visit **www.lawpack.co.uk** or call **020 7394 4040**

MORE PRODUCTS AVAILABLE FROM LAWPACK

Book-Keeping Made Easy

Many businesses fail in their first year or two because of insufficient financial control. This guide provides the new business owner with an understanding of the fundamental principles of book-keeping. Includes procedures for the sole proprietor and small business, accounting for growing businesses, double-entry book-keeping, ledgers, payroll and final accounts.

Code B516 | ISBN 1 904053 85 8 | Paperback | 153 x 234mm | 104pp | £10.99 | 2nd edition

Business Letters Made Easy

Business Letters Made Easy provides an invaluable source of more than 100 ready-drafted, annotated letters to take away the headache and time-wasting of letter writing. The book covers managing suppliers and customers, debt collection, credit control, employing people, sales and marketing management, banking, insurance and property, business and the community and international trade.

Code B520 | ISBN 1 904053 87 4 | Paperback | 153 x 234mm | 280pp | £12.99 | 1st edition

Employment Law Made Easy

Written by an employment law solicitor, *Employment Law Made Easy* is a comprehensive, reader-friendly source of information that will provide answers to practically all your employment law questions. Essential knowledge for employers and employees. Valid for use throughout the UK.

Code B502 | ISBN 1 904053 69 6 | Paperback | 153 x 234mm | 224pp | £11.99 | 5th edition

To order, visit **www.lawpack.co.uk** or call **020 7394 4040**

MORE PRODUCTS AVAILABLE FROM **LAWPACK**

MORE PRODUCTS AVAILABLE FROM **LAWPACK**

Employment Contracts Kit

This Kit contains what an employer needs in order to prepare contracts for staff and so comply with legal requirements. Full- and part-time, temporary and domestic contracts are included, with a manual that discusses the relevant areas of employment law. A free CD with contracts and recruitment and management letters is included. For use in England & Wales and Scotland.

Code P110 | ISBN 1 904053 18 1 | Sealed Wallet | 305 x 220mm | £9.99 | 2nd edition

Limited Company Kit

This Kit explains what a limited company is and provides the documents needed to set one up with a registered office in England, Wales or Scotland. It includes copies of the necessary Companies House forms, Memoranda of Association, Articles of Association, Share Certificates and an instruction manual that guides the reader step-by-step through the process.

Code P101 | ISBN 1 904053 22 X | Sealed Wallet | 305 x 220mm | £9.99 | 5th edition

Self-Employment Kit

Going self-employed is a major decision in life. In effect, you need to meet the demands of running a business yourself. This Kit provides a framework on which to prepare yourself and to develop your own accounting systems. It includes a wealth of practical advice and provides template documents for cashflow, budgets and financial control. For use in England & Wales and Scotland.

Code P115 | ISBN 1 904053 07 6 | Sealed Wallet | 305 x 220mm | £9.99 | 1st edition

To order, visit **www.lawpack.co.uk** or call **020 7394 4040**